T0326869

WHAT A MUSHROOM LIVES FOR

WHAT A MUSHROOM LIVES FOR

Matsutake and the Worlds They Make

Michael J. Hathaway

With a foreword by
Anna Lowenhaupt Tsing

PRINCETON UNIVERSITY PRESS

PRINCETON AND OXFORD

Published by Princeton University Press
41 William Street, Princeton, New Jersey 08540
99 Banbury Road, Oxford OX2 6JX

press.princeton.edu

All Rights Reserved

Library of Congress Cataloging-in-Publication Data

Names: Hathaway, Michael J., author.
Title: What a mushroom lives for : matsutake and the worlds they make / Michael J. Hathaway.
Description: Princeton : Princeton University Press, 2022. | Includes bibliographical references and index.
Identifiers: LCCN 2021034507 (print) | LCCN 2021034508 (ebook) | ISBN 9780691225883 (hardcover) | ISBN 9780691225890 (ebook)
Subjects: LCSH: Tricholoma matsutake. | Tricholoma matsutake—Ecology.
Classification: LCC QK629.T73 H38 2022 (print) | LCC QK629.T73 (ebook) | DDC 579.6—dc23
LC record available at https://lccn.loc.gov/2021034507
LC ebook record available at https://lccn.loc.gov/2021034508

British Library Cataloging-in-Publication Data is available

Editorial: Fred Appel and James Collier
Production Editorial: Ellen Foos
Text Design: Karl Spurzem
Jacket art and design: Karl Spurzem
Production: Erin Suydam
Publicity: Sarah Henning-Stout and Kathryn Stevens
Copyeditor: Kathleen Kageff

This book has been composed in Arno Pro with Futura Display

Printed on acid-free paper. ∞

Printed in the United States of America

10 9 8 7 6 5 4 3 2 1

And the mushroom hunters walk the ways they walk
and watch the world, and see what they observe.
And some of them would thrive and lick their lips,
While others clutched their stomachs and expired.
So laws are made and handed down on what is safe. Formulate.

The tools we make to build our lives:
our clothes, our food, our path home . . .
all these things we base on observation,
on experiment, on measurement, on truth.

—Neil Gaiman, "The Mushroom Hunters"

CONTENTS

FOREWORD

Fungi! Delicious, dangerous . . . daring? In Michael Hathaway's hands, mushrooms come to life not just as objects of human fears and desires but also as makers of more-than-human worlds. Hathaway is curious about the *longue durée* as well as the contemporary. Why did mammals evolve warm-bloodedness many millions of years ago? How do present-day Tibetans in Yunnan raise funds to revive traditional architecture? These and many more strange and wonderful things about the world turn out to rest on the actions and reactions of fungi.

When Michael Hathaway, Tim Choy, Lieba Faier, Miyako Inoue, Shiho Satsuka, and I formed the Matsutake Worlds Research Group together, fungi were hardly interesting to either scholars or the general public. Perhaps it was the sense that we had stumbled on something fascinating but really obscure that allowed us to completely underestimate the size of the project on which we had embarked. A few months of research, we thought, and perhaps the publication of an edited volume. Many years later, we're still at it—but in an entirely different context. Our group has not only produced an edited volume (*Matsutake Worlds*[1]) but we are also two volumes into an expected trilogy of single-author matsutake ethnographies. (In addition to Hathaway's book, see Anna Tsing, *The Mushroom at the End of the World*,[2] and Shiho Satsuka, in preparation.) Meanwhile, the climate for reading and writing about fungi has radically changed: now everyone is interested, from chefs to ecologists to public intellectuals.

There are many streams flowing into this river of mushroom fever, and readers will get to explore them in *What a Mushroom Lives For*. Here are three that matter a lot to me: First, natural history has come alive again in the past decade as a mode by which both scholars and ordinary people connect with the diversity and fragility of the world around us. Suddenly the sex lives of tube worms and the communicative practices of birds hold public attention.

Enlightenment thinkers, we now recognize, underestimated nonhumans in making them appear to be passive resources for human free will. Fungi now greet us from within a newly enlivened and socially active world of nonhumans. This world sparks endless curiosity. Perhaps we are making up for lost time in getting to know the creatures around us.

Second, fungal vitality enters and pushes forward a revolution in biology. Throughout the twentieth century, the "modern synthesis"—the coupling of genetics and evolution—held the attention of biologists. Working from the imagined lives of mammals, biologists saw autonomous organisms as representatives of autonomous species. Relations with other species were those of eat or be eaten. Today, this view looks archaic and ideological: interspecies relations of all sorts had been ignored. It turns out that most organisms cannot develop into themselves without working with other species. We humans cannot digest our food without gut bacteria, for example. Fungi are in the thick of this scientific transformation, particularly those fungi involved in mutualistic relations with plants. These fungi bring water and nutrients to plants in exchange for carbohydrate meals. Some scientists think that the only reason trees can form forests is the support they get from mutualistic fungi. Thinking through interspecies relations changes the game entirely for ecology, evolutionary theory, and developmental biology. Evolution, for example, is no longer imagined as a species-by-species affair. "Holobionts" are multispecies evolutionary units: they evolve together.

Attention to interactions across species also brings history into the heart of ecological attention; interactions are always contingent and historical. History, in turn, opens the door for new forms of interdisciplinary interaction, and this is my third stream. If our object of study is ecological history, we need humanists as well as biologists to trace just how things came out this way. After several hundred years of academic contempt between the humanities and the natural sciences, new avenues of collaboration are becoming visible. Again, fungi have been at the heart of this opening. Fungal histories are a most promising arena for such work across disciplinary divides. Michael Hathaway's sketch of the interplay among humans, barley, yaks, oak scrub, and matsutake in Tibetan Yunnan is an example of what becomes possible in this new conversation.

What does a mushroom live for? The first half of Michael Hathaway's book presents material on the social lives of fungi, with *and* without human companions. The latter, that is, the attention to social relations even where no humans are present, is a radical break with much thinking in the social sciences. The field has indeed been coming to terms with the need to integrate nonhumans into analyses of the sociality of humans—but not social relations between one nonhuman and another nonhuman. For many humanistic social scientists, leaving humans out of the scene of sociality raises the specter of a non-reflexive positivism. Anthropologists in particular have spent many years distancing ourselves from this kind of positivism, which can reduce both humans and non-humans to passive objects of a God-like gaze. Hathaway's insistence that mushrooms make worlds by associating with insects, soil minerals, and tree roots—as well as sometimes humans—pushes the critique of positivism in a new and provocative direction. To me, it is useful to think with science-studies scholar Karen Barad's "intra-action," that is, the social encounters through which beings take on form. Fungi *become* mutualists in particular interactions with roots; in other interactions with the same tree's roots, that same fungal species may become a parasite. (Matsutake is one example of a fungus that can take both these forms.) These are versions of Barad's "agential cut," through which worldly form emerges in social relations. However, this is only one alternative to positivism. Hathaway's book should inspire a lively discussion of theories of social action in which humans may or may not be part of the scene. As Hathaway explains, even considering the expensive, large, and aromatic fruiting bodies of matsutake, most of the mushrooms are never found by humans. And yet they too are active in making worlds.

The second half of Hathaway's book takes us to southwestern China and to the histories of humans as well as nonhumans in which mat-sutake mushrooms play a part. While I have never had the chance to visit the Tibetan areas of Yunnan, I accompanied Hathaway for two field seasons in the Yi area described in this book, and I benefited a great deal from his longer research in the area. Seeing the photograph of our down-to-earth friend Li Bo (figure 5.1) brings me back to the pleasure of getting to know people in Nanhua. Nevertheless, Hathaway turns the experience on its head by pushing us to consider how the town is produced by the social activity of mushrooms as much as the other way around. Mushrooms, he argues, are explorers, relationship builders, and performers, creating landscape and more-than-human community. The emergence of cultural venues dedicated to presenting Yi ethnic culture, for example, builds on the ways the mushroom requires place-based knowledge, which in turn has built ties among Yi buyers and pickers. Meanwhile, far from dismissing nonhuman companions, pickers work together with insects in seeking out mushrooms, whose aroma draws each into previously unexplored niches.

Similarly, in Hathaway's discussion of the Tibetan region, humans are only one of many active players in making more-than-human worlds. The interplay between the loss of grassland for yak grazing, on the one hand, and the rise of prices for matsutake, on the other, stimulated energetic matsutake picking among humans. Perhaps human interactions with yaks and matsutake could have supported each other, but the overlap between the season yaks need intensive attention and the season matsutake fruit made this impossible. Hathaway's account requires readers to embrace multiple nonhuman protagonists. The story makes no sense without them.

Do mushrooms make the world? You may think that sounds like a science fiction proposition. Not so. Read this book to explore this question in the world around us.

—*Anna Tsing*

PREFACE

Fungi are perhaps the most unappreciated, undervalued, and
unexplained organisms on earth.

**—Chenglei Wang, "Do Not Ignore the Role of Fungi
in Biodiversity Conservation," 2014**

Fungi are world-makers: powerful subjects that shape our planet in
largely unrecognized ways. This book explores this claim from the perspective of cultural anthropology, the study of culture, rather than mycology, the science of fungi. When I began this project, my goal was to
conduct conventional anthropological research centered almost solely
on human lives. To that end, I carried out fieldwork in Southwest China's Himalayan forests, where I explored the changes that took place
within two mountain communities—one comprising ethnic Tibetans,
the other ethnic Yi—after the people living there started to sell a valuable wild mushroom, called matsutake,[1] on the global market. With its
distinctively fresh, sharp smell, cinnamon-like and peppery, matsutake
has become one of the most distinctive signifiers of Japanese national
identity and thus is highly valued by Japanese consumers. I witnessed
how this new economy, which began in the 1980s, significantly altered
the lives and livelihoods of Yi and Tibetans. Among other things, the
infusion of previously unimaginable sums of money reconfigured the
overall position of these particular groups within China's traditional
ethnic hierarchies; it also altered intergenerational relations and re-
shaped gendered patterns of work within each of these ethnic groups.
In one of these communities, ethnic Tibetans are now building massive
new homes with matsutake cash, and some families are getting rid of

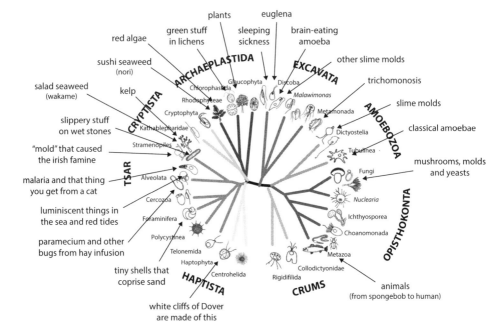

FIGURE P.1. The family tree of life. Animals and fungi are shown in the lower right. Figure adapted from Adl et al. 2012. With kind permission from Anna Novák Vanclová.

their yak herds, severing a relationship that goes back thousands of years. In the other community, ethnic Yi are funneling a significant stream of funds into cultural revitalization, encouraging the creation and consumption of Yi-identified foods, song, and dance. I watched this new economy emerge as a community-led phenomenon, one not organized or imposed by the state, as has frequently been the case in China.

In the course of doing this anthropological fieldwork, I also spent time with Chinese and Japanese matsutake scientists in their laboratories and field stations, and thus my project began to expand. Hearing about their exciting research into matsutake's mysterious form of life piqued my own curiosity. This, in turn, led to my obsession with all things fungal, and I began delving into the scientific literature on the strange and wonderful abilities of our close kin (I say this because animals and fungi belong to the same evolutionary group known as opisthokonts).[2] As I later heard from mycologist Lawrence Millman, plants are our second cousins, while other animals and fungi are our first cousins, which you can see on a recent vision of the tree of life (fig. P.1),

modified in a more layperson-friendly manner by the evolutionary pro-
tistologist Anna Novák Vanclová. I have long been an avid reader of ecology, but I came to realize that new insights into the global importance of
fungi in shaping our planet challenged some of the fundamental biological principles that I had been taught. This realization motivated me to
write this book, which builds on a wide range of recent scientific findings
to present a case for fungi as absolutely critical to the development and
continuity of life on Earth as we know it.

I embarked on this journey with my fellow collaborators—Tim Choy,
Lieba Faier, Miyako Inoue, Shiho Satsuka, and Anna Tsing—collectively
known as the Matsutake Worlds Research Group (MWRG). For more
than a decade, we have been talking, traveling, and sometimes writing
together to create special journal issues.[3] We conducted joint anthropological fieldwork in the United States, Japan, China, and Finland, as we
followed the matsutake from remote mountainsides to high-tech labs
and to high-end grocers and dinner plates. Since the 1980s, the matsutake mushroom has become the object of a global hunt by more than
half a million people. These quiet hunters walk slowly, poking under
pine needles and oak leaves, searching for this elusive mushroom. According to some matsutake experts, almost every single high-quality
matsutake is destined for Japan, making it the center of a global trade
network that can amount to more than several billion dollars a year. In
Japan, matsutake are considered the most prized of all mushrooms.
They were once a sumptuary good: that is, only royalty were allowed to
eat them. Later, they entered commerce, and in the last century, people
have paid as much as $2,000 a kilogram for matsutake. The price varies
widely from day to day and year to year with lows of $20 a kilogram, and
there are different grades with significant differences in value. Although
Japan was self-sufficient in matsutake until the 1940s, afterward their
numbers started to decline because of complex environmental changes,
creating what was deemed a national tragedy. Until the 1970s, domestic

production fulfilled Japan's needs, but by 2012, it met only 3 percent of domestic demand. In the 1970s, Japan started building a global importing network that by 2012 was supplying 97 percent of the national market.[4] Of all the regions, Southwest China's Yunnan Province became the most important matsutake exporter to Japan (fig. P.2).

I was excited to have another excuse to carry out research in Yunnan, the place that first introduced me to a realm where mushrooms were well loved and well known. When I first began working in Yunnan in the fall of 1995, I was amazed at the abundance and prominence of mushrooms for sale. They seemed to be everywhere: in baskets along the roads, in piles at the farmers' market, and wrapped in plastic in the grocery store—even the airport offered a dozen varieties of dried mushrooms in expensive boxes. I had come from California, where my fascination with mushroom culture had just begun; at that time, finding wild, edible mushrooms was still considered a bit eccentric and mainly the preoccupation of elderly Italian men and middle-aged hippies. In China, however, the woods were thick with mushroom hunters, tracking down something to sell or to cook for dinner. During my time in Yunnan, I taught at a forestry school and made lasting friendships with those who would continue to guide me on my return trips.

I went back to Yunnan from 2000 to 2002 to carry out my doctoral fieldwork on international environmentalism; I returned again in 2005 after I joined the Matsutake Worlds Research Group. For the next decade, I returned on numerous occasions during the summer or fall, and I stayed with communities of ethnic Tibetans and Yi. I went into the forest with them to hunt matsutake. I spent mornings drinking perpetually refilled mugs of salty butter tea and nights sipping *baijiu* (hard liquor) while *laoban* (bosses) captivated their motley crews of mushroom buyers with high-stakes gambling and tales of bravado. In these mountain villages, the gap between the rundown houses of the past and the new "matsutake mansions" was remarkable. Groups of older women sometimes lamented the ways this newfound wealth was turning things upside down, as some men lost their fortunes in gambling while others sold their yak herds and bought trucks to haul goods or vans to go into the "underground" taxi business.

On trips with Chinese colleagues or with other members of the MWRG, I visited research centers in China and Japan. I traveled from high-altitude forests near glaciers to bustling markets in Yunnan's capital and then eventually flew alongside boxes of matsutake to Japan, where consumers eagerly awaited them.

This book is the second installment in one of the world's first academic trilogies written by a collaborative team, and you can see what happens when a group of anthropologists is infected by a powerful curiosity about a globe-trotting mushroom and its interlocutors. Some of you have read the first installment in our trilogy: Anna Tsing's book *The Mushroom at the End of the World: On the Possibility of Life in Capitalist Ruins.* You don't need to read the trilogy in order, as I don't assume any knowledge from Tsing's book, but I highly recommend it. Tsing asks us to imagine how we might engage the world when one of the key assumptions generated by the Enlightenment period—the idea of progress—can no longer be taken as an article of faith. As Tsing argues, our capitalist economy and even our expectations of history were built on an assumption of continual progress, but such a faith seems to hold true less and less. Extending Tsing's inquiry, this book asks how, then, do we continue to build livable worlds, not only for ourselves, as humans, but also for our more-than-human cousins on whom we rely but who also have their own intrinsic right to exist? Put somewhat differently, I am interested in questioning the notion that was fundamental to the Enlightenment, but has earlier precedents, that humans are separate from the natural world—different from and superior to other animals—a notion that can be labeled "human exceptionalism" or "human supremacy."

FIGURE P.2. Map showing Yunnan Province in regional context. Expansion of Yunnan showing main field sites near Gyalthang/ Zhongdian and Chuxiong, in relation to Kunming, the provincial capital. Map by Saki Murotani.

To some degree, this kind of questioning is supported by some of my fellow anthropologists who conduct what they call "multispecies ethnography" to explore the lives of other organisms.[5] Beginning in the early 2000s, they started to shift away from a human-focused account of the world to ponder how people live with other animals—without viewing the latter as mere resources for humans. This work is rich and fascinating, and yet I also realize that most of it tends to keep humans at the center of the story. In contrast, I want to explore the lives of fungi, both with and without human engagement, in a world where humans are not always the main protagonists and actors.

I explore these lives through the concept of "world-making." In one sense, the meaning of this term seems obvious, for don't all living beings carry out their lives through activities that shape the world around them? Yet I use this term deliberately, to challenge the common notion that only humans are active world-makers and that all other organisms simply react to stimuli. In contrast, world-making suggests that all living organisms constantly *interpret and engage* their surroundings, and thus *creatively participate* in making their worlds. A number of anthropologists have used the term "worlding" as a way to actively describe human activities, looking at how particular cultural groups carry out their own life projects. Here, I extend this idea of actively making worlds to go beyond humans, including other animals, plants, and, not least, fungi. As I will later explain, I pluralize the term "world" to explore the ways in which each organism, to some degree, perceives and engages with a distinctly limited universe, and yet this universe is also shared with others.

In this book, I show how fungi are actively sensing, choosing, and acting in their lives—and I demonstrate how important they are in making the wider world we share with them. Different species of fungi have different capabilities and predilections, but as a kingdom they possess some fairly unique capacities. My intention, however, is neither to argue for "fungal exceptionalism" nor to suggest that fungi are the most important forms of life. Rather, I wish to explore the particular world-making dynamics that fungi carry out. My initial interest in matsutake is a result of my involvement with the MWRG and because of its importance in areas of the world where I've lived and conducted research; later, my interests expanded to the kingdom fungi because I wanted to explore the larger implications of fungi for life as we know it. Such an inquiry could, in fact, be usefully extended to *any* organism, whether another kind of fungi such as lion's mane or shiitake, or a member of any kingdom of life. We could examine the world-making capacities of worms, bats, or bacteria. A handful of accounts describe some active qualities of certain organisms (such as Charles Darwin's accounts of worms and microbiologist Lynn Margulis's accounts of bacteria), but these are few and far between, as well as different from what I describe here.

For those of you who don't yet know much about fungi, I hope this book might ignite your curiosity. When I began this project, the "mushroom renaissance" had barely started, but by now I have met many of the people who have inspired my work and taught me a great deal. For those of you already under the spell of fungi, this book offers insights into fungi's diverse abilities to transform the world, and I hope to provide more respect and less idealization, as well as to challenge the predilection toward turning fungi into the carbon-sequestering, plastic-eating solution to human-made woes. For those of you who are already suspicious of widespread claims of human exceptionalism, this book might fuel your thinking and foster new ways of being. For those of you practicing ethics-centered diets and forms of consumption such as vegetarianism or veganism, who draw a clear line of demarcation between the capacities of plants and animals (and typically put mushrooms in the plant realm), I hope this book will persuade you to rethink this division.

My own journey has been transformative. I was raised to fear fungi, but after years of interacting with them—as student, observer, eater, grower, and experimenter—I now see them as critical life partners for all forms of life, not only for the plants that they create symbiotic relations with, but also for humans. I hope to convince you that fungi, which are a kind of organism that many people can't imagine acting in lively, interactive ways with its fellow beings, are in fact doing just that. Focusing on these interactions points to the ways that, in fact, every organism is actively making worlds all around us, worlds that we share, whether we notice them or not. This happens in relation to and despite the current tragedy of climate change, cascading extinctions, and increasing pollution. To experience life in which humans are not at the center encourages a newfound attention and curiosity about how we make lives together. It is my hope that through such a perspective we might move toward a reenchantment of the world that would see other beings as just as alive and complex as ourselves.

WHAT A MUSHROOM LIVES FOR

Introduction

In Southwest China's Yunnan Province, high on the Tibetan Plateau, you can climb a hillside at dawn during the late summer and early fall, and looking down, although the night is ending rather than beginning, it will seem as if stars are coming out in the valley below.

They are flashlights, carried by villagers walking up into the mountains to hunt for matsutake mushrooms. People collect all morning and return home when the dealers arrive at a village market or drive along the roads, buying from mushroom hunters as they go. These dealers will, in turn, sell their matsutake to other dealers, until eventually the mushrooms reach Japan, where they are highly valued.

The mushrooms are carried in shoulder satchels, often hand fashioned from bags once used for fertilizer or pig food, the durable everyday sack of rural China. When hunters find a prize specimen, they wrap it in a layer of thin plastic film or snap off the tip of a rhododendron branch and wrap grass around the mushroom, securing it to the branch and cinching it snugly, like a swaddled baby. The mushrooms are carefully cushioned in these satchels, as they can be easily damaged during the long hike across steep terrain.

Over the season, millions of these mushrooms travel from forests to villages, to local buying centers and bulking stations and then onward to Japan. Because of their delicacy and their great appeal to insects, such a journey must be made within forty-eight hours, for the insects within the stalk and the cap are already starting to eat them.

—Revised field notes from Yunnan Province

This book reveals a world that is far more lively, far more unexpectedly propelled by multispecies relations than the human-centered world that most of us learned about in school. Since the 1980s, in both popular and

FIGURE I.1. Young matsutake cushioned in lichen and wrapped in paper.
Found in the mountains outside of Gyalthang, in Yunnan's main Tibetan area.
Photo by Michael J. Hathaway.

academic literature, there has been an explosion of interest in lives "be-yond the human." Many people now actively learn about and debate whether animals possess rights, have capacities to feel pain, or imagine the future.[1] Yet most of this intellectual energy has been focused on the lives of our fellow mammals with familiar bodies and behaviors. Many of these studies take a "sameness approach," showing how other animals are similar to humans.[2] This book, in contrast, takes us beyond the ani-mal realm to explore a place barely known to most people, the inner realm of fungi and how they participate in making the world around them via their relations to microbes, other fungi, plants, and animals. Even when I describe social worlds in which human mushroom hunters in China could be regarded as the main actors—seeking out matsutake and bringing them to lucrative markets (fig. I.1)—I also show how the mushroom's own behaviors shape these human actions in ways that are rarely considered. All told, most matsutake appear and disappear without ever entering into human markets, yet they nonetheless shape landscapes and, hence, human lives, in powerful ways.

The book tells some stories of how the kingdom of fungi has shaped our planet in powerful ways, and it asks how we might imagine a different history of Earth. I tell tales of the intimate relations between fungi and other beings that many have not heard before but that nevertheless deeply influence our lives. Although I first turned to anthropologists and biologists to help me understand how other organisms might be understood as actors (i.e., as having the capacity *to act* rather than simply react), I quickly discovered that there are, as well, a growing number of other people—including fiction writers, historians, and social scientists, among others—who are no longer satisfied imagining a world where humans are the only beings that seem to act.[3] They write novels that acknowledge other species' lives and create histories in which humans are not the only actors that matter, with all other species relegated to the background as merely potential resources, commodities, or threats to human world-making ventures. They consider social lives as not just limited to the social lives of humans. Although such challenges to human-centered accounts of the world might seem new, they are not at all; indeed, a sense of society that includes other beings has long been common and remains strong in most parts of the world, despite the strong tendencies in Western thinking to view humans as exceptional beings.

Biologists are one group of people who are explicitly interested in the lives of nonhuman organisms, and they have produced much of the intimate knowledge we have about these beings. Science, as we understand it today, as a formalized and very particular way of thinking about and describing the world, is relatively new. It was only in 1833 that the term "scientist" was coined, while the older term "natural philosophers" slowly lost favor. In the 1860s, Charles Darwin's radical theories, which argue for our common kinship with other beings, were sometimes described as a way to bridge what was known as the "Great Divide," meaning the gulf that separates humans and other animals. The Great Divide is a form of "human exceptionalism," the notion that humans are categorically distinct from all other animals. Another related concept, the "Great

Chain of Being," asserted that humans not only are different from other animals but are better: smarter and more powerful. This concept ranked all beings, from God to angels, humans to animals, and plants to minerals, along a hierarchical scale (lords above peasants, lions above zebras, gold above lead, and so forth). Given this, Darwin's theories encountered great resistance, especially from the Christian Church; nevertheless, in time these theories reigned supreme among scientists.

Like all forms of knowledge, our understandings of biology and its categories were shaped by the social conventions in which they were formed. For instance, the biological sciences still use terms invented in eighteenth-century Europe that depict the world as mirroring the dominant political structure of that time: as "kingdoms" of plants, animals, or fungi.

Another concept from this time, which once seemed unassailable but is now increasingly debated, is that of "species."[4] Previously, scientists held that organisms belonged to the same species if they could mate and produce offspring that in turn could produce offspring; the most frequent example given in biology books was that while a horse and a donkey could reproduce, their offspring, a mule, is sterile, thus demonstrating that a horse is a different species than a donkey. Such a "test" may work fairly well for large mammals. This definition, however, has now been challenged as we have come to understand that much of the biological diversity of life is microscopic. We now know that microbes can share genetics with each other "horizontally," that is, not only "vertically" from parent to child but from sibling to sibling and even between organisms that are considered different species.

Even the term "animal" serves to reinforce a sense of human exceptionalism when it is used to describe all animals *apart from* humans. There are some recent alternatives that remind us that we, too, are animals, but these often rely on unwieldy phrases such as "nonhuman animals," "other-than-human animals," or "other animals." In this book, I use the term "nonhumans," even though I dislike language that continues to perpetuate human exceptionalism. Others are trying to create new words to manage this "lexical gap in the English language."[5] One example is "anymal," a nonanthropocentric term to describe *any* animal other than one's own species. Thus, from the point of view of a garter snake, a blue jay is an anymal.[6] I could have used the term "anymal" in this book, but I am often talking about nonhumans that are members of other kingdoms, such as plants and fungi.

While this may seem to be merely a matter of semantics, animal rights advocates such as Kenneth Shapiro have been insightful critics of such linguistic distinctions between humans and animals as a kind of "categorical error" that creates a "subordinate and inferior status for nonhuman animals."[7] To repeat these errors, then, is to reinscribe this Great Divide—a divide that I believe has had and continues to have disastrous consequences.

The strength of the conviction that humans stand alone, in that they alone have minds, is in part a legacy of ideas enshrined by René Descartes (1596–1650) and others during the Enlightenment.[8] They argued that the best way to understand how nature works is via a mechanistic model—and indeed, the most prominent metaphor of the time was a complex yet familiar machine: the clock—consisting of hundreds of tiny parts that function in predictable ways. Descartes and his allies argued that animals were like small machines: their actions were predictable responses to external stimuli.

This view was not totally dominant, however, even within Europe. Charles Darwin, for example, conducted experiments in which he approached other organisms such as worms as lively beings who possessed forms of mysterious intelligence and who learned from their experiences.[9] Nonetheless, an increasing number of scientists began to use a mechanistic framework.[10] In a long battle that took place over centuries, two groups began to emerge. One group saw organisms as agentive, *self-making* machines (and indeed, the word "organism" refers to the idea that the animal was self-organized as opposed to being a "mere machine"). The other group saw organisms as more passive, as subject to their environment. By the 1930s, however, this battle had basically been won by the more passive-minded group, later termed the "neo-Darwinists."[11] Today, nearly a century later, the conceptual framework of the field of biology continues to be almost exclusively a mechanistic one. Although in everyday life, many biologists appreciate the organisms they work with as lively beings, the constraints on scientific writing tend to produce accounts described in mechanistic terms. As Eileen Crist shows, standard biological texts use mechanistic language to portray "animals as objects . . . [who] appear blind to the meaning and significance of their activities and interactions. . . . Behavior comes through as something that happens to an animal, rather than an active accomplishment."[12]

This mechanistic understanding has been incredibly productive for scientific discovery in several ways. For example, viewing animals as machine-like meant that they were perceived as not feeling pain or possessing rights; hence, moral questions could be disregarded, and a wide array of experiments—from limb amputation to vivisection—could be carried out on them. The goal of a mechanistic view is to discover the mechanisms of life, the relations of cause and effect. In fact the term "the mechanism" is a common way to describe such connections. The main query of a mechanistic framework is always, at base, *What is the mechanism?* This question, while useful in many cases, also excludes many other lines of inquiry by, on the one hand, assuming a mechanism both exists and controls each relation, and, on the other, by assuming that the question of a mechanism is in and of itself the most important question.

Yet, as I will show, like all frameworks, a mechanist model also constrains the kinds of questions asked, the experiments designed, and the conclusions drawn. Experimenting with a livelier framework enables different inquiries and new understandings.

Indeed, a number of scientists are now challenging some of these precepts, suggesting potential problems with mechanistic approaches and arguing that nonhuman animals (as well as plants and other organisms) engage their surroundings in a dynamic way. They are also questioning the notion of human exceptionalism. Many claims have been made, such as the claim that only humans can use tools, feel pain, imagine the future and their own death, and communicate through language. I was imbued with this belief system whether or not I wanted to be, and even when I had a sense that it wasn't always right, I had neither the vocabulary nor examples to challenge it. Only recently, as I explain in

chapter 3, have I learned how certain language conventions work to con-
tinuously de-animate the world, drawing a strict demarcation between
humans and other species and in turn rendering most of the nonhuman
world passive.

Other scientists, rather than assuming a Great Divide, are now arguing that many animals do, indeed, use tools, feel pain, and communicate
in sophisticated ways. It is no longer shocking news for a scientist to
argue that elephants have emotions, that bees communicate through
intricate dances in the hive, or that dogs mediate the emotional lives of
their owners.[13] Biologists, who once insisted on the machine-like quality of animals, are now revealing in intimate detail the fantastic abilities
of our fellow animals. In tandem, my own thinking is enriched and be-
holden to a number of scholars in the social sciences and humanities
who have taken a keen interest in the more-than-human world.[14] Together, their work helps show us how the "social" is never that of humans
alone. These are exciting times.

This book brings us from the more common focus on animal lives to a
focus on organisms far less familiar—fungi. In it, I aim to show you that
our very world has been made possible by the daily actions of trillions of
fungi that have shaped our planet for almost a billion years. The discipline
of biology, deeply shaped by a particular European legacy, has tended to
be animal centric, and this orientation has markedly limited our ability to
notice, understand, or appreciate the great diversity of nonanimal forms
of life, or to imagine how organisms such as fungi have influenced animals
and ecosystems, how they are part of world-making dynamics.

What are fungi? Biologically they compose an extraordinarily diverse
realm, and most are microscopic. Within a European context, fungi

were often thought to be "strange plants."[15] It was only as recently as 1969 that fungi were recognized by scientists as a separate kingdom from plants, and yet the institutional structures are so deeply entrenched that now, even more than fifty years later, almost all professional mycologists were trained in a botany department.

Let me provide you with a brief sketch of fungal lives. They are diverse and superlative in a number of ways. Although many fungi are microscopic, it's possible to walk up the steep hills of Oregon's Blue Mountains for hours and still be traveling on the body of one of the world's largest known living organisms, an *Armillaria* fungus also known as a honey mushroom. Over its vast life, possibly more than two thousand years, this fungus has created a mycelial network extending over two thousand acres (eight hundred hectares). Although they seem not to move (except by mysteriously appearing overnight), hidden within their bodies is the capacity for unbelievable movement: many fungi shoot out spores with incredible acceleration, far greater than astronauts face in a rocket ship. But, because they are so tiny, they face a lot of air resistance, and thus attain speeds of only up to seventy miles per hour (113 kilometers per hour), a nonetheless impressive mark that ties with that of the cheetah, the world's fastest land animal.[16]

Fungi are like animals in several ways, for indeed we are both within the larger category I mentioned earlier: opisthokonts. This means we each need to consume other organisms to survive (rather than being able to photosynthesize like plants or eat chemicals like some bacteria). Moreover, we both breathe in oxygen and exhale carbon dioxide. Some fungi eat live animals, hunting them with snare traps and sticky nets. Many form intimate relations with animals, traveling within or on their bodies, including our own. Some fungi live within insects that transport them to new locations inside of trees, where the insects lay eggs; the fungi grow so that when the eggs hatch the young insects are surrounded by food—fungi-digested wood that is nutritionally enhanced.

While most fungi are microscopic, others are so massive you could seek shelter under them in a rainstorm (fig. I.2). The microscopic fungi are all around us, and many of them we intentionally eat with gusto: yeast that transforms wheat into bread, barley into beer, and grapes into

FIGURE I.2. A range of fungal forms, not to scale. From top left: beer with yeast; chytrid
spores swimming after frogs; mold growing on bread and sporulating; matsutake; lichens
growing on a tombstone; chaga growing on birch. Image by Saki Murotani.

wine. Fungi, thus, bring bread into being, but they can also take it away
from us, as many people keep bread until blue mold appears in ever
expanding circles. If you look closely, lovely stalks appear as the mold
starts to produce spores to find new homes. Other molds seem to appear overnight: a blemish on an orange's skin expands and slowly liqui-
fies its insides. Some fungi last for centuries or more. Next time you're
at a cemetery, study the lichens growing on tombstones, as mycologist
Anne Pringle does. While the stones represent human death, they may
offer lichens a site for potential immortality.[17]

Likely the world's most common representation of a fungus is the classic *Amanita muscaria* found in the book *Alice in Wonderland* and the video game *Super Mario*, with its distinctive red cap with white flecks. Many large mushrooms appear in the autumn, popping out of the ground in a range of colors and shapes; some call them the "jewels of the forest," or "flowers of the fall." Some look like a dead person's fingers poking through the earth. When they get large, they can become incredibly charismatic: one giant cluster of *Tricholoma giganteum* mushrooms in Yunnan attracted more than 130,000 visitors in just a few days, many of whom threw money at them.[18]

To human noses, mushrooms emit a bewildering range of smells, but relatively few people have experienced their diversity of odors. The "classic mushroomy smell" of the most popular mushroom in the West (*Agaricus*) is only the beginning:

> *Lactarius hibbardae* smells like coconuts, many *Inocybe* species have a spermatic odor, *Cortinarius vulpinus* has the odor of a sow in heat; *Trametes suaveolens* and the aniseseed polypore (*Haploporus odorus*) have the odor of anise; chanterelles smell like peaches or apricots; *Amanita citrina* smells like raw potatoes; *Cortinarius paleaceus* smells like geraniums; *Mycena alcalina* smells like Clorox bleach and *Russula xeromphalina* smells like cooked crab.[19]

We know little about what effects these smells have on other species: who can detect them, who finds them attractive, and who finds them repellent? We recently learned that some insects may be attracted to glowing fungi, and other fungi possess skin with stunning ultraviolet colors, invisible to humans but likely capturing the attention of other beings.

Fungal sexuality has often been an enigma, and for many years, fungi were seen by European researchers as a kind of mysterious plant. While technically called a "fruiting body," a mushroom is not a fruit that contains seeds. It is closer to a flower that contains pollen yet is very different as well. Like a flower, a mushroom is a sex organ. Plant flowers disperse pollen, which can be carried by wind, rain, insect, or animal, and when these grains land on the receptive part of a flower's ovary (sometimes called a stigma), the pollen germinates, creating a long tube that then ejects sperm into the ovary; if the sperm and ovary fuse, they create a seed. In fungi, on the other hand, spores do not immediately seek a companion of a different "mating type." Historically, European systems of understanding often assumed that there were two mating types in the world, what we often call "the sexes," a supposedly universal division between female and male, construed as opposites. Study of the sex lives of fungi (and bacteria), however, helped foster a new understanding of diverse mating types. Through the work of two female mycologists— Elsie Wakefield, who directed the world's largest mycological department at Kew Gardens in England, and Cardy Raper, who worked at Harvard—breakthroughs were made in understanding fungal sex and the proliferation of mating types. They discovered that spores are choosy about which spore they will mate with and will not fuse with those of the same type. Although some fungi have a few mating types, others can have tens, hundreds, or even thousands. One kind of shelf fungus boasts more than twenty-three thousand mating types, a mind-boggling notion, as many humans still puzzle over reckoning with any more than two genders. Such profligacy means that when these spores germinate, say on a new tree, the chances that they will find spores of a different type are greater than 99 percent.[20]

Yet the aboveground mushroom is just a tip of a fungal iceberg. Underground, the structure of a fungus is formed by many filaments, called hyphae, that grow together in a mycelium. Just one cell thick, hyphae extend through the soil column in search of water and nutrients. Some fungi are saprobic, meaning they eat dead matter such as fallen leaves and wood, breaking it down into soil and food for other organisms. Some fungi are symbiotic, meaning they make intimate relations with other living organisms, in a wide array of dynamics, from mutualism to parasitism. One form of mutualism—the connections between fungi and plant roots in what is called a mycorrhizal relationship—has become one of the planet's most fundamental relationships, because fungi are critical to the flourishing of nearly 95 percent of the plants around us. I will explain this more in detail in chapter 2, but basically fungi are able to gather water and soil nutrients, which they exchange for photosynthetic sugars created by their plant partners. Thus, almost every plant is part of an underground fungal network.

Fungi deviate from the usual dichotomy of plant and animal; fungi are the "third *F*"—what some call the "funga," alongside flora and fauna. Their ways of living diverge significantly from other organisms; for example, some fungi make symbiotic connections with species in other kingdoms, while others produce a prodigious diversity of sexes. Some fungi may have a potentially unlimited life span, for they do not possess what scientists call "preprogrammed cell death"; that is, although they get older, they don't, as we commonly understand it, "get old." To acknowledge fungi's seeming strangeness, according to animal and plant logics, can help us queer the field of biology, revealing a radically different view of life.

Within mainstream Western education, fungi have been largely ignored, sometimes described as the "forgotten kingdom." They have tremendous powers over our lives, and yet most accounts of the world leave them out entirely. Even in the late 1990s, botanists argued that plants were often underappreciated compared to animals and came up with estimates that of all living things on Earth, plants made up 99 percent of living biomass.[21] This figure, however, was derived from a "two kingdom view of life" that excluded fungi.

Newer studies of the weight of the living world are more likely to consider fungi. New estimates suggest that they weigh more than humans and all other animals on the planet combined.[22] Even biology textbooks tend to give them short shrift, vastly understating how fungi are shaping the world. Such neglect even influences our main models of climate change, to our detriment, given that mycologist Jennifer Talbot recently revealed that while all the major climate change predictions completely ignore fungi, they are likely the most important actors in determining how carbon moves through soils—a source of carbon ten times more important than all land-based human processes.[23]

Fungi have been such influential world-makers in part because of their success in learning how to live in diverse habitats, from high mountaintops to the ocean deep. Even in one small area, such as between your toes, scientists have found more than forty species of fungi that have made a home.[24] They have spread throughout the entire planet by water, land, and air, including places such as the ice fields of Antarctica, where plants cannot survive.[25] Recently, fungi were found deep underground, within the igneous oceanic crust itself, previously thought to be without life.[26] Their lightweight spores, designed to be airborne, allow fungi to traverse vast distances in space. Although we tend to think of fungi as stationary, like how we understand plants as rooted in place, fungal spores can move over space, flying high in the jet stream, crossing oceans, and jumping continents.[27] In general, though, mushrooms move by spore or mycelia with plants, expanding or shrinking their territories slowly in relation to climatic changes, moving ahead of or behind glaciers. They have been thriving on the planet for so long that many have moved around the world riding on tectonic plates.

Each kind of fungi has its own particular history of global movement, the study of which is called "mycogeography." Let's consider one possible map of the matsutake diaspora over tens of millions of years, which contains a sense of matsutake's movement. Whereas most maps convey an impression of spatial fixity and temporal stability, this one offers a dynamic representation (fig. I.3). The map was developed by Britt Bunyard in 2013, and it portrays three different species, each represented in a different shade (some of these species were later subdivided, in 2017).[28] Notice how the lightest path shows this particular species traveling across the vast Atlantic Ocean, which gives a sense of matsutake's long-distance travels, though as I explain in chapter 4, when these journeys happened, so long ago, the arrangement of the continents was very different than it is now, so perhaps the Atlantic Ocean did not yet exist.

Even as we view this map of matsutake's travels, it's important to keep in mind that the global spread and distribution of matsutake was by no means assured in advance; each encounter was dynamic, and how each unfolded was not a foregone conclusion. Rather, their expansion into new lands—and unfamiliar ecosystems—meant that there were many encounters with organisms that were new to them. Matsutake became involved in a whole suite of relations: some were able to connect with

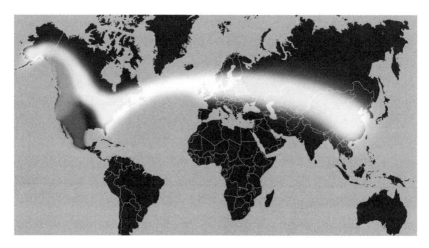

FIGURE I.3. A map showing the possible spread of matsutake species.
With kind permission from Andreas Voitk.

certain unfamiliar trees, were parasitized by some plants that could detect their underground mycelia, and were eaten by other fungi and animals. It should be noted that organisms that tried to eat the newly arrived matsutake had to experiment. Although some fungi have strong flavors that either attract or repel, some seem edible but contain toxins that take days to bring death—and yet somehow this knowledge spread, albeit imperfectly.[29]

I wanted to understand how such relations emerge, but it is rare to find scientific accounts of mushrooms as dynamic organisms. The cases that are most likely to show fungi as dynamic are those that label them "invasive," when they kill plants or animals or objects we care about, such as the fungi that kill bats or frogs, those that attack human crops, or the fungi (dry rot and others) that devastate wooden ships and homes. In almost every other case, scientific accounts portray fungi as relatively passive, as already adapted to their environment—in short, as mechanistic organisms in a mechanistic world. Many scientific accounts portray fungal life as a mere series of chemical reactions, each triggered by a certain stimulus. And still, despite these reductive portrayals, scientists fall in love with fungi, experiencing their charisma while spending long hours peering into microscopes or digging down into the forest duff to observe their mycelial networks.[30]

Working with my anthropologist colleagues in the Matsutake Worlds Research Group, I had initially envisioned this book as a more conventional anthropological account in which people were the only actors, but something happened along the way. In part, I became fascinated on my trips to Southwest China by the ways that two neighboring ethnic groups of people, Tibetan and Yi, were committing themselves to the trade in this valuable mushroom.

On the eastern edge of the Chinese Himalayas, previously poor and poorly treated rural villagers of two different ethnic groups transformed their lives with mushroom wealth, but in notably different ways. Their encounters with matsutake brought about rapid social changes that went beyond a mere infusion of wealth. I was originally interested in what people *did* with matsutake and, more specifically, how it was a commodity that people *used* to make social worlds. I was more familiar with the social role of labor in making money, but less so with the idea that different kinds of labor made different kinds of wealth. For example, I was intrigued when Tibetans told me that "matsutake money" was not the same as "yak money." I found this quite interesting, but it was two particular interactions over the course of my research that expanded my interests into the capacity of the mushroom itself, beyond its role as a commodity in a human-made world, and helped me to start imagining the various ways that it, too, could be understood as a world-maker.

At a lecture in Texas four years ago, someone asked me if it was possible for Chinese pickers to pick every single matsutake, so that there would be none left. I replied that the life cycle of matsutake was starkly different from the "animal model" most biologists used to predict sustainable levels of harvest.[31] For example, picking a mushroom is not like killing a deer or a salmon because the mushroom is not the entire organism. As I explain above, mushrooms contain not seeds but spores, and I offered the audience member the most common analogy to describe the whole fungus as being like an apple tree: the mushroom is the aboveground apple, and the root-like group of hyphae is the belowground tree. Picking apples once a year doesn't reduce the crop the next year; that is, it doesn't affect the tree's sustainability.

I had given these answers before, but this was the most extreme version of the question I had ever gotten. It made me realize that the vast majority of matsutake would never be picked by people because they grow over such a vast area, on steep and challenging terrain, and often hidden under leaves. Pondering my answer led me to this question: How might I understand the lives of the many mushrooms that never become part of human commodity chains?

A second encounter compelled me down another path of understanding how matsutake *themselves* are world-makers. My first main introduction to this idea came from some ethnic Yi matsutake-hunting families who described insects as savvy and active agents who were hunters of mushrooms, like themselves. I joined these matsutake hunters on morning hunts and to evening concerts of lively music with exuberant crowds and tipsy musicians. They described how certain matsutake-hunting insects could detect the mushrooms by studying the actions of other insects who were also able to find hidden matsutake. In other words, they were describing these insects, such as the fungus gnat (*Sciaridae*) or click beetle (*Elateridae*), as actively perceiving and reflecting on the world. I had never imagined that insects might study the actions of other insects that were neither prey nor predator. The Yi matsutake hunters carried out experiments to see how these insect matsutake hunters engaged the world, which I describe in chapter 5. This understanding of insects as capable, intelligent, and able to learn felt very different from how I had been taught to understand them, that is, as purely instinctual or as animals without brains.

Back in Vancouver, I read deeply on the topic of insect-mushroom relations and came across an intriguing term: "biosemiotics." It seemed to resonate with Yi sensibilities. In graduate school, I had learned about "semiotics," the creation and interpretation of human-made signs as a form of communication, including writing, speech, body language, and so forth. I had not heard that there were researchers exploring the semiotics of other organisms in an expanded field called "biosemiotics." I had heard a few stories of animal communication such as how the alarm calls of vervet monkeys distinguish between various threats—from a snake, eagle, or leopard;[32] but biosemioticians were exploring many organisms in varied and sophisticated ways. Even though I could imagine that mushrooms might send out signals to other mushrooms (as I had read that maple trees send chemical alarm messages to other trees when attacked by insects), I realized I held a relatively impoverished view of the possibilities of fungal communication.[33]

Biosemiotic accounts helped reorient my point of view in two fundamental ways. First, I had previously assumed that such reactions were merely automatic reflexes, based on the mechanistic language of Western science that asserts that certain responses are triggered by particular stimuli. In contrast, biosemiotics introduced me to the notion that organisms exercise discretion and choice in how they respond. Second, I was surprised to learn that mushrooms not only produce forms of chemical communication; they also *interpret* others' communication. Initially I had assumed that communication was a simple case of cause and effect, in which a particular situation (such as an insect attack) would trigger the automatic release of a corresponding chemical. In turn, others of the same species would be triggered by perceiving that corresponding chemical. Such chemical messages, however, are not just one-to-one, where each situation is matched with a corresponding chemical. Rather, the receiver needs to interpret these messages. I wasn't as surprised that mammals have this interpretative ability, but I didn't imagine mushrooms could also engage the world in this way. I realized, to my chagrin, that in regarding mushrooms as relatively passive organisms, I was more subject than I had thought to some of the mechanistic frameworks that imagined other living beings as machines reacting to their environment. These two chance encounters exemplify how we are remade through our interactions with others and led me to dig deeper into the lifeworld of matsutake and other fungi.

I increasingly became aware that matsutake were not just a pawn of human economic projects; rather, and much more radically, they were carrying out their own life projects. Through these two encounters, my research project turned from being a human-centered account—in which humans are the only world-makers, who use other organisms as objects to accomplish their human worlds—to an account of how matsutake seek out other species to make their own worlds. This book explores how the actions of matsutake themselves—as living, breathing organisms interacting with other fungi and other organisms—literally make the forest and knit it together. My initial fascination with this one fungus grew into an interest in the whole kingdom and in the ways that fungi have shaped the planet.

Agency, Ontology, Actant

Above, I have used the concept of world-making to show how matsutake literally and actively make the world. The notion of world-making allows us to think in more nuanced ways about a related concept: *agency*. Within mainstream Western understandings, agency is often defined as *intentional action*[34]—the corollary being that only humans carry out intentional actions and thus that *agency* is exclusively a human domain. *Intentionality*, in turn, typically presumes an autonomous individual whose actions are consciously willed.[35] That said, I wish to sidestep debates about intentionality, because I am interested in how actions create worldly effects rather than the internal motivation for such actions.

Some scholars of nonhumans argue that the concept of agency is too historically burdened and anthropocentric to use beyond a human context.[36] But I contend that the notion of agency has too powerful a grip on the imagination, and on the fundamental structure of scientific thought, to confine it to humans. Inventing a separate term to describe nonhuman agency merely reinforces the idea that only humans have agency, thereby supporting a belief in human exceptionalism. In contrast, I am interested in finding ways to think about agency that include but also extend beyond humans.

Over the past few decades, various thinkers have introduced concepts that help us understand agency in new and different ways. The French intellectual Bruno Latour is the most well-known thinker and promoter of the influential framework of Actor Network Theory, which relies most notably on the concept of the "actant."[37] With this term, Latour meant to dissolve the distinction between humans and nonhumans (whereby actors can be *only* humans), arguing that we should see humans as just one actant among many.[38] Likewise, American political scientist Jane Bennett's notion of "vibrant matter" suggests a kind of neo-animism that recognizes the ways that nonhuman and nonliving things may possess something like agency in the world.[39] Some describe Bennett's work as having changed the dominant Western understanding of things as a passive backdrop for human affairs into an understanding of things as active protagonists in and of themselves.[40] I find Bennett's work exciting and inspiring; it argues that mainstream political theory largely views the world as consisting of dead matter, as a collection of objects that can be manipulated for human purposes. In contrast, she argues that the agency of things often exceeds human intentions, not only in episodic crises, such when a giant electrical grid fails, but in the comings and goings of everyday life, when piles of leaves clog street drains.

Compared to the terms offered by Latour or Bennett, one of the main advantages of the term *world-making* is that it inspires our curiosity into the particular qualities, properties, and lifeways of specific living beings. Whereas Latour's and Bennett's understandings tend to flatten agency, so that all things are equally actants or forms of vibrant matter, world-making inquires into the particulars—the diverse and specific agencies—of various organisms. For instance, African elephants' world-making will be totally different from the world-making of termites they share lands with, or even substantially different from that of their closer kin, the Asian elephants.

World-making as a concept appeals to me in part because it challenges the notion of human exceptionalism and can accommodate all organisms insofar as it explores how particular beings act in the world and create relationships with others. In contrast to the Enlightenment perspective that only humans deliberately make worlds, I argue that all organisms—be they plant, animal, fungus, bacteria, or otherwise—are world-makers, even if their world-making occurs in ways that differ radically from each other (including from humans). The German philosopher Martin Heidegger stated that only humans make worlds, that animals are poor in world, and rocks have no world.[41] Although I appreciate that he does not make an absolute distinction between humans and other animals, I disagree with his assumption that all other animals are poor in world. If we avoid the assumption that other animals are not as sophisticated as humans, new areas for biological research suddenly open up. What kinds of new questions can be asked? What else can be noticed? For example, later I will pose the questions: How might a matsutake-loving fly observe and learn from other insects when searching for mushrooms? How might fungi themselves learn or gain skills throughout their lifetime?

My understanding of the term "world-making" draws on several earlier, related terms that have been used within cultural studies and anthropology. In these fields, world-making has sometimes been used to discuss the practices and perspectives of particular cultural groups. In cultural studies, Lauren Berlant and Michael Warner provide an example that I found helpful to my thinking, describing what they call "queer world-making" as efforts to build convivial networks and create amenable spaces within homophobic cities.[42] In anthropology, the closest term is often "cosmologies" or "ontologies" (which, unlike in cultural studies, typically refer to a group of people with its own distinct language), where each group is understood as living in its own world. As Anna Tsing argues, studies of cosmologies or ontologies have most often used these terms to mean a *mental* conceptual space, whereas the notion of world-making also includes physical activities and emphasizes the active making of a place, rather than assuming that it is always already there.[43] People have asked me if I will write this book in an ontological way, that is, from matsutake's point of view or "how mushrooms think." Although I am interested in this topic, I do not presume to have any such insights on it, nor have I seen any scientific studies claiming such knowledge. Instead, I am interested in understanding how matsutake *act* in the world.

Scholars of ontology—which briefly means a theory about the nature of being or what it means to have existence—suggest that each cultural group lives in its own world, which necessarily presents the idea of plural worlds. This idea, in turn, challenges the dominant understanding that we all live in a singular world—a *uni*-verse. To articulate this concept, several of these scholars of ontology (who I will call "ontologists") have proposed the notion of a "world of many worlds" or a "*pluri*verse."[44] They use the term *pluriverse* to challenge the hegemony of the Western scientific perspective of a singular truth that describes a singular "nature"—a universe. Their work seems to parallel one of the defining contributions of anthropology, which is, as Ruth Benedict put it in 1934, to pluralize the notion of culture. Benedict and others suggest that in contrast to the European idea of Culture, there are many different ways of being (cultures). These anthropologists helped us to recognize forms of Eurocentric thinking that permeated how other groups were evaluated and judged. Like Benedict and others, ontologists argue these other cultures should not be judged and ranked on a Eurocentric scale of more or less civilized. Yet, ontologists have sometimes criticized their fellow anthropologists for holding other cultural worlds at arm's length (dismissing them as "myths," "legends," and "beliefs") and conflating the world of their own culture as *the* world; ontologists, on the other hand, take other cultural worlds seriously in and of themselves, as fully rich worlds that are each unique.

Although I appreciate ontologists' notion of the pluriverse, I nevertheless want to expand its use from the diversity of *human* worlds to include the worlds of all beings.[45] Expanding the notion of the pluriverse to include other species seems especially apropos given that these scholars of ontology were often deeply influenced by Indigenous groups in the Andes and the Amazon, whose important philosophical understandings recognize and acknowledge how other beings are important actors.[46]

A second point of divergence is that I am leery of how some more extreme versions of ontology imagine the world as a realm of social bubbles. To put it another way, anthropologist Eric Wolf famously critiqued scholars who describe cultural encounters as if each group were like a billiard ball, preformed and solid.[47] As Tsing argues, in contrast to thinkers who write about cosmologies or ontologies as solid forms that "collide" or "clash," the concept of world-making allows us to imagine such worlds as overlapping and intersecting.[48] I find it helpful to follow what Lieba Faier (a MWRG colleague) and Lisa Rofel describe as an encounter approach:

> Ethnographies of encounter distinguish themselves by considering how culture making occurs through everyday encounters among members of two or more groups with different cultural backgrounds and unequally positioned stakes in their relationships. The term encounter refers to engagements across difference: a chance meeting, a sensory exchange, an extended confrontation, a passionate tryst. Encounters prompt unexpected responses and improvised actions, as well as long-term negotiations with unforeseen outcomes, including both violence and love. Ethnographies of encounter focus on the cross-cultural and relational dynamics of these processes. They highlight how meanings, identities, objects, and subjectivities emerge through unequal relationships involving people and things that may at first glance be understood as distinct.[49]

Although this statement was made in relation to human-based encounters, the concept of encounter is also highly applicable to interspecies relations. As American scholar Donna Haraway puts it: "human becoming is becoming with" other species.[50] Haraway looks at human-dog relations where dog domestication was previously viewed as a one-way street: humans took a wild dog and turned it into a docile domesticate. In contrast, an encounter approach that views dogs and humans as "becoming with" encourages the further study of how dogs may affect human evolution, as well, such as shared emotional responses (including producing oxytocin) and shared viruses. Viewing domestication as a long-term encounter, a mutual transformation, means that we have to look at *both* organisms as actors. Trying to specify these ancient changes is hard if encounters do not leave physical traces; I was intrigued, for example, to learn that some people in Japan have a unique enzyme that allows them to better digest seaweed, and the theory is that this enzyme is the result of a multispecies encounter.[51] I first assumed that this enzyme must have emerged from human DNA but later read that it was created by bacteria in the gut microbiota. Scientists think this enzyme became part of some people's gut microbiota through a horizontal transfer of seaweed-eating bacteria a long time ago. Thus, rather than being digested in the stomach, these bacteria become part of human bodies. This is a powerful example of the physical manifestations that particular encounters can leave on the body. Although world-making is shaped by particular bodies, the outcome of encounters is uncertain. The presence of one life inflects the life of another, in complex webs of relations that are in the process of becoming.

I was reminded of these encounters when one day I found a huge chicken-of-the-woods mushroom, a bright-yellow-and-orange shelf fungus that I always found delicious. Grilling it over coals after a brief marinade, I was surprised by its intense bitterness, which I belatedly realized was because it had grown on a willow tree and had thus imbibed the tree's salicylic acid (a bitter chemical used in making aspirin); previously I had found these mushrooms growing on oak trees, and together they produced a flavor that I found delicious. Although these mushrooms had a similar shape, color, and texture, each was inflected by its encounter with other beings, in this case a tree, and became something different. In a less human-centered example, mycorrhizae are a unique structure that occurs only when fungal mycelia and plant roots find each other and decide to fuse together. Even though there are two kinds of membranes in between, I consider this a form of fusion as these connections allow for the exchange of water and many nutrients across these membranes. Mycorrhizae exist, in other words, only through encounters.

Even if one takes the plurality of worlds metaphorically rather than literally, world-making suggests that different species partake of their own significant and unique experience of living, and one can refer to each of these qualitatively different experiences as a kind of world. In this way of understanding, neighboring species experience realms that may be largely unknown to each other even as their presence shapes the other. For example, the sensorial worlds of perception and realms of interest for two species in the same place—say, for example, an African elephant and a mosquito on his or her ear—can be extremely different, yet in other ways their lives coincide as they are briefly connected when the elephant's blood enters the mosquito's body.

When this project started to expand and I began to explore whether fungi had any active shaping effect on the world—that is, whether they possessed any kind of agency—I was hard-pressed to imagine that they could. I wondered: How did their presence matter to the world? For a while, my main answer was that they were a source of food for the animals who ate them. Even as food, however, they seemed inconsequential: low in calories, temporally fleeting, and scattered throughout the landscape. I later realized that I was thinking only about the kind of forest mushrooms I knew best and my own relationship to them, which meant the limited timescale of their aboveground lives. Breaking from these limitations, I switched my perspective to explore the role of fungi writ large, comprising more than a million species, to learn about their actions under the soil, in seawater, within the bodies of logs and the leaves of plants. My time frame shifted from a few weeks to hundreds of millions of years. In changing scale, forms of fungal agency loomed large, so much so that I soon discovered that fungi are absolutely critical to the planet as we know it.

But then, too, I also wanted to learn about the specific capacities of one particular fungus, the matsutake. There was relatively little research about its specific ecological role, in part because humans' as yet unfulfilled desire to cultivate it has focused research on how it grows. Matsutake's role within the larger ecology is also complex because it grows alongside hundreds, perhaps thousands, of other fungal species, so it would be incredibly difficult to disentangle the effects of matsutake in particular within a given ecosystem. Scientists have eliminated large mammals, such as deer, from a given landscape in order to study how their absence affects other animals and plants; but to date, no scientist has attempted to eliminate matsutake from a given site to see what hap-

pens to their coinhabitants. Such a task would be formidable, as matsutake exist as underground mycelia without fruiting in many places, so
their presence is not easily detectable. Likewise, their underground
growth is extensive and may run deep into the soil, so it would be very
difficult, nearly impossible, to remove just this one species of fungus.[52]
I hope, through future experiments and by intimately working with
matsutake, we can better understand its specific engagements with
other beings and how it shapes the larger forest ecology.

I have gathered what ecological knowledge I could find on matsutake,
and by taking a lively, world-making approach, I have emphasized how
they have shaped their own experience of life, as well as the lives of
many others, including people, with whom they are in intimate relations. In the spirit of encounter, I became fascinated by how this one
particular fungus, the matsutake, shapes its encounters with different
groups of people. Toward that aim, I traveled to Southwest China,
where I had previously spent nearly three years, to see how the two main
groups of matsutake hunters have made their lives together with this
mushroom and other beings. Together, fungal studies and anthropological fieldwork combined to make this book possible.

Chapter 1 provides an account of the history of planetary life, with
fungi in the starring role, as told through the stories of fossils and molecular DNA. I ask, What role have fungi played in turning the blue
planet into the green planet, and thereby in making this larger world?
This chapter explores fungi as an entire kingdom to ask how, over deep
time, fungi have shaped our planet, allowing life to leave the oceans to
make a living on land. In so doing, they stopped the clock on the creation of fossil fuels and played a crucial role as mediators among water,
rocks, and plants.

Chapter 2 explores the ways that fungi seek each other out, along
with nutrients, water, and plants, and how they engage with insects and
mammals. The chapter highlights the ways that even without direct
human presence, fungi negotiate and build their own networks, and it
explores the range of their often-surprising capacities.

Chapter 3 shows different ways that some Western scientists have
tried to explore the inner worlds of various species. I introduce the reader

to the work of the polymath Jakob von Uexküll and consider his inspiring legacy in fostering curiosity about how different animals, including insects, perceive the world and draw meaning from it. The work of Uexküll and others has been critical to creating a lively approach to engaging with other beings as fellow perceivers and interpreters of the world.

In the second half of the book, I turn to the matsutake and explore how neighboring but culturally distinct ethnic groups draw this valuable mushroom into their lives in different ways. Chapter 4 tells the story of Japan's mushroom tragedy, when matsutake nearly disappeared, and how the Japanese turned to China and elsewhere to find new supplies of matsutake. I show how Southwest China became the world's largest source of matsutake, through geological events that made matsutake lives possible on the edge of the tropics. I also look at the daily actions of hundreds of thousands of matsutake hunters and dealers who bring the mushrooms by foot, bicycle, car, truck, and bus to airports, and then fly them to Japan.

Chapter 5 moves us into Yi villages, revealing how the matsutake economy does not take a single form, even among neighboring groups. I look at how the Yi people in China's Yunnan Province learned about Japan's desire for matsutake and articulated themselves into transnational markets. I explore how the Yi learned to attune themselves to matsutake's rhythms and ways of life, including its conditions of growth, and how they share the hunt for matsutake with a range of insects. Yi are making their own worlds, using the mushrooms to do so, yet the matsutake's own particular way of life shapes how the Yi build worlds around them.

Chapter 6 is about Tibetan mushroom hunters in Yunnan Province, for whom matsutake wealth has created an unprecedented reverse diaspora. A number of Tibetans who had fled to India are now returning, in part attracted by the thriving economy generated by matsutake, which includes a housing boom of "matsutake mansions" and a resurgence in the creation of Buddhist paintings and carvings. I explore how two of the most prominent species with which Tibetans in this region have entangled their lives—yaks and matsutake—converge and diverge in challenging ways. Yaks and mushrooms are not just pliable objects of human projects or ambitions; they each have their own demands and desires, and the worlds they cocreate are shaped by these continual encounters.

In my conclusion, I offer some thoughts on the implications of taking a world-making approach. How, for example, might this mode of inquiry challenge some of the deeply embedded human exceptionalism that is structured within English and many other languages, as well as the field of biology?

Eventually, I learned that the vast majority of matsutake are never picked by humans: they quickly rise and fall above ground, attracting a wide variety of insects, birds, and mammals to their fruiting bodies. As well, below ground, they live their mycelial lives shuttling nutrients and water between trees and the realms of soil, minerals, and water; they spend the bulk of their time slowly traveling through subterranean depths over decades, perhaps centuries, helping to determine the kinds of trees and the forest ecology that they help produce. Even for those mushrooms that are picked by people, we can see how the Japanese demand for freshness—another term for alive and still breathing[53]—means that the mushroom's ways of life continue to exert a powerful force as a range of ethnic groups rush it from forests to markets.

By using a world-making perspective, we can understand matsutake as active beings that possess capacities of perception and action; the idea of world-making also allows us to explore how these organisms seek out and create networked relationships across different kingdoms of life. As a kingdom, the worlds of fungi may go far beyond our animal-centered imagination as well as our typical scientific assumptions. It is my hope that this project, which builds on the efforts of many others, plays some role in encouraging a shift in our understanding of how we live with other beings in complex and interdependent relations. To explore how a mushroom—typically seen as possessing no capacity for awareness or choice—interprets and acts in making a world can in turn open up a greater awareness of a lively planet, where humans are far from being the only actors that matter.

CHAPTER ONE

Fungal Planet

The Little-Known Story of How Fungi Helped Foster Terrestrial Life

In 2017, I spent four months as a visiting professor at the University of Hawai'i. There, I learned about an earth that was far more dynamic and alive than I had previously imagined. I saw new land being made as streams of lava poured into the ocean, starting to harden almost instantaneously as it hissed and roared. Making land in this way was a slow process, as these waters were deep. Yet one could see vast stretches of land that rose from the ocean, where lava had been slowly added, layer by layer, until at last it was exposed to the sun. Then, the land began to erode and fall back to the sea—slowly at first, through pounding rains and powerful winds; sometimes more suddenly by massive landslides that in turn created tsunamis that smashed into neighboring islands leaving boulders perched well over a hundred feet high.

I enrolled in a university class to learn more about this strange and wonderful land. The professor asked how long it would take for life to start transforming this landscape of new lava rock, with its thin, hard skin, and glowing molten insides. "Several years," answered one student. "Less than a year," said another. The professor then answered herself: "Almost instantaneously: microorganisms are floating everywhere, and some will soon adhere to this new earth, surviving the heat and making their niche." We had all been contemplating macroscopic animals like spiders, forgetting about the microscopic beings like bacteria and fungi that make our worlds possible.

Later, hiking on the Big Island, I walked over vast stretches of new lava flows, improbable landscapes of hard rock that grew unbearably hot in the sun. My only referent for these rocky places made by fire were mountaintops made by ice along the high ranges of North America's west coast. Glaciers traveled over the region ten thousand years ago, bulldozing forests and soils in front of them and, in retreat, leaving behind stretches of bare rock, polished and scratched. On the mainland, forests slowly returned to these soil-scraped lands, but some places remained bare. In Hawai'i, however, I discovered another timescale, where fresh lava became thick forests in just a few centuries or less. I experienced a fluid, rapidly changing earth, where land is created and lost almost instantly, not transformed and polished over millennia. I could imagine the possibilities of ancient Earth at the start of terrestrial life, how organisms landed on hard lava and turned it into soft soil, creating a place for plant and animal life. At the time, I had no idea these first organisms were likely fungi.

—Reflections on fieldwork

In 2007, Alan Weisman wrote the now-famous book *The World without Us*, a well-informed, speculative account that began with a simple yet powerful question: What would happen to Earth if humans vanished? Weisman shows there would be all kinds of disasters if humans disappeared, including the destruction of more than four hundred nuclear power plants, yet surprisingly, even with thousands of catastrophes, let alone the breakdown of our oil pipelines and so forth, he suggests that life on Earth would quickly return if humans disappeared. Considering this scenario, I couldn't help but ask my own version of the question by adding a "fung" to Weisman's query: What would happen to "the world without fungus"?

What if all fungi disappeared from planet Earth? This world would be profoundly different, and the result would be catastrophic. Fungi play a major role in decomposition, that is, breaking down the dead bodies of plants and animals. Without fungi, masses of dead grass, herbs, and trees would pile up and up, likely blocking the growth of the next generation. Not only would this take up land, but, as importantly, the disappearance of fungi would mean that many of the planet's nutrient and carbon cycles would either grind to a halt or be so severely slowed that some plants and animals would not have enough nutrients to keep growing. Thus, as death piled up, life could not renew itself. As scientists are increasingly learning, however, fungi not only play a crucial role as "decomposers" or "rotters" that break down life; they also are absolutely essential to many forms of life coming into being, especially plants. As I will later explain in more detail, they do this through connections between plant roots and fungal bodies. Fungi are living within and around the roots of nearly every plant on the planet, and plants have relied on fungi's assistance for hundreds of millions of years. If fungi disappeared, we would face a plant apocalypse, as plants would no longer get enough water to drink and food to eat. Indeed, for most plants, more nutrients are provided by fungi than by their own roots.[1]

It is extremely difficult to imagine our planet without any fungi, especially as they are so diverse and omnipresent. Yet on the flip side, they are often unseen and relatively little studied. Fungi are all around us and also within us, but few of us know much or anything about what they do and how they affect the world. For some time, I tried to imagine the world without the one kind of fungus that I am most interested in, the matsutake, but I could not find enough scientific knowledge about their particular ecological role to envision the effects of its absence. What would happen to forests without matsutake's presence? It turns out that while matsutake need certain trees to grow, we don't know of the opposite reliance: there aren't any known trees that require, in particular, matsutake's assistance. Many animals eat matsutake, but I couldn't find any that ate *only* this mushroom—so its disappearance, I assumed, would likely not cause any known visible species to starve. I later learned that this was not completely true, for there is one plant, the candy cane (*Allotropa*), that seems to rely completely on matsutake, so if matsutake disappeared overnight, *Allotropa* likely would too. But again, there was little information about *Allotropa*'s ecological role, so it was difficult to envision the effects of its absence. This led me back to my original thought experiment about what would happen if all fungi disappeared, but this time I approached it from the opposite direction: not based on a speculative future where they disappeared, but on a possible past where they had never evolved.

Taking this tack, I wondered what might happen if the history of life on planet Earth was looked at, not through the usual focus on animals, but on fungi as the central players. How might this history of life differ?[2] First, such a shift would reveal some of the critical relationships that make our world possible yet are little known. Second, and perhaps most importantly, a fungi-centric view might reframe some of the dominant underlying frameworks that biologists use to explain life. For instance, accounts of life tend to focus on animal-centered forms of competition between individuals of the same species (competing for mates, scarce food, and habitat for activities like nesting) or different species (in terms of predator-prey relations).[3] Such relatively simple and antagonistic relations are stressed in "survival of the fittest" accounts. This notion of life, which builds on a Victorian view of nature as "red in tooth and claw," was first articulated as a scientific claim around the time of Darwin's main writings in the 1860s.[4] More than a century later, such perspectives remain strong; I clearly recall my own indoctrination into such perspectives via TV documentaries in the 1980s, when the announcer, David Attenborough, described, in his authoritative older-male British voice, dramatic scenes as a "ruthless battle for survival."[5] These programs encouraged a vision of nature that was overwhelmingly competitive, a vision that is now increasingly being challenged by a few ecologists,[6] especially those who have worked closely with organisms that have prominent mutualisms with other species, such as bacteria[7] and truffles,[8] and are recognizable even in the intimate relations between animals and bacteria writ large.[9] In my fungi-centric account, I emphasize the importance of complex relationships that are interspecies, and even interkingdom, some of which are antagonistic but many of which are not.

Turning to macroscopic fungi, basic assumptions about the nature of their ways of being are now coming into question (fig. 1.1). For years, it was thought that each fungus fell into one of three categories: saprobic, parasitic, and mutualistic (mycorrhizal). Saprobes eat dead bodies,

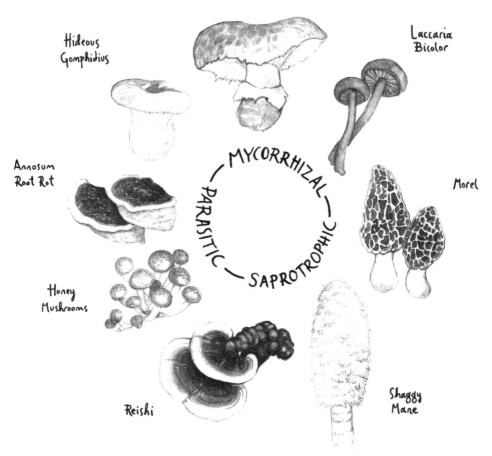

Matsutake

Hideous
Gomphidius

Laccaria
Bicolor

Annosum
Root Rot

Morel

MYCORRHIZAL

PARASITIC

SAPROTROPHIC

Honey
Mushrooms

Reishi

Shaggy
Mane

FIGURE 1.1. The three main categories of macroscopic fungi as a continuum. From Arevalo 2019. Image courtesy of the artist, Carmen Olson, and New Society Publishers.

whereas parasites eat living bodies, and mutualists create mutual benefits with the living. These were thought to be permanent qualities, fixed types of relationships between fungi and plants or animals, yet scientists found that some fungi can dramatically change their own modes of living when circumstances change. As early as 1925, researchers began to notice that some fungi were mutualists in the field, but under laboratory conditions, they became saprobes.[10] And in 1963, Kazuyoshi Hiromoto found the matsutake in the field could change from being a mutualist to a saprobe within a single lifetime.[11] Thus, these relations are not always fixed but can be dynamic.

Fungi are powerful world-makers. Here, I describe how the world was in large part made possible by fungi and how fungi continue to shape it. In crafting a history of the fungi-generated world, I focus on four important ways fungi shaped the world as well as hint at relationships that are only recently being explored. First, starting about a billion years ago, fungi were critical in chemically breaking down rock and creating the beginnings of soil. Second, fungi, through their underground hyphae, formed intimate relationships with almost every single plant species, hence enabling plants to survive and flourish. Third, fungi invented ways to digest wood and turn dead plant bodies back into food for others, rather than having these former plants be locked away in an underground tomb of nutrients (becoming what we call "oil" and "coal"). Fourth, fungi were powerful parasites and predators: their ability to harm and kill bacteria, plants, and animals made them major players in shaping the diversity of life.

The ubiquity of fungi in every known terrestrial ecosystem is, in fact, a legacy of the ancient history of these organisms and a powerful testament to their ability to adapt. In addition, many scientists were surprised to discover recently that fungi are also plentiful in the oceans; scientists had imagined that fungi were terrestrial and that the ocean's salinity was hostile to fungal life.[12] As I will soon explain, however, a long time ago, fungi were found only in the sea, and they allowed the first movement of life from sea to land. As plants and animals diversified, so did fungi, and it is hard to imagine any plant or animal without relations to fungi. Such fungal relations with plants and animals are far more diverse and complex than many realize. These relationships, in fact, are largely unknown or ignored, hidden from virtually everyone except for a handful of some of the world's most dedicated mycophiles. I bring you into this world of fungal life and into the worlds that mushrooms are making.

Before doing so, however, allow me to introduce a brief caveat, which is that for the next few pages I will be knowingly mycocentric. I recognize that the world is shaped by a diverse set of mainly microscopic organisms (often referred to as "microbes"), not only fungi but also bacteria, viruses, and others, a whole slew that makes and keeps our planet vibrant. Unfortunately, much of the scientific and lay orientation toward microbes has focused on them as pathogens, as "germs." This pathogenic bias toward microbes, however, is changing, in part through stories like those told in this book, which reveal how thriving life on Earth depends on microbial organisms and their complex interrelationships. Yet, as the global Covid-19 pandemic makes clear, some fear of microbes is reasonable as humans remain deeply vulnerable to viruses and other microbes. The pandemic has done a great deal to undermine a sense of human mastery and shows a virus's power not only to attack and kill humans, but to compel people to reconfigure the global economy and nearly all aspects of social life to a degree unprecedented for a century (since the so-called Spanish flu epidemic of 1919).[13]

Fungal World-Making: How Fungi Made the Planet Green

In terms of our present-day planetary ecology, fungi have been and remain key world-makers. Fungi, in fact, were critical to developing the complex ecology enjoyed today. Fungi have, for hundreds of millions of years, been changing the planet into one that supports more than just oceanic life, one where plants and animals could live on land. Although fungi did not directly enable animal life, as far as we know they enabled land plants, which, in turn, allowed land animals to flourish. In terms of the history of Earth, the importance of fungi goes back in deep time to when fungi were significant forms of ocean life. Indeed, for billions of years life existed only in the ocean, but finally some species started to explore the land.

Turning Rocks into Minerals

For more than half a billion years, terrestrial fungi helped make the rocky earth inhabitable for plants, in part by helping them to weather rock and break it down into assimilable pieces.[14] Fungi, along with many other organisms, were key to creating soil. For a long time, soil scientists thought that fungi played only an insignificant role in "rock weathering" (the process by which solid rock breaks into smaller pieces and eventually into minerals available as plant nutrients), less than 1 percent of the effect compared to water and wind.[15] However, re-searchers now suggest that fungi, along with other microorganisms, play a significant role in breaking up rocks, creating mineral nutrients for plants' nutrition as well as for their own.[16] Fungi's exploratory threads—called hyphae—are a strange paradox: soft, delicate, and fragile, each strand is easily broken, yet over time they convert solid rocks into slurry. These slow actions, repeated by untold trillions of hyphae and accumu-lated over millions of years, are key to making soil.

Fungi's active role in breaking up rocks was only recently discovered by Antoine G. Jongmans and his colleagues, in 1997.[17] They were puz-zled to find that even in European forests suffering from acid rain, plant growth was still strong.[18] Somehow, plants were still acquiring nutrients that technically should not have been available, because normally the soil would have been considered too acidic and thus the nutrients would have been "locked up." Jongmans's team discovered that fungi provided trees with these otherwise unavailable nutrients. Fungi are often de-scribed as exuding digestive acids that dissolve surrounding minerals, which are then "absorbed" by the fungi—the same passive term used for the relationship between a paper towel and a water spill. However, when Jongmans's team looked at soil under a microscope, they found grains of sand with tubular holes.[19] The team was puzzled and at first suspected the holes were due to atmospheric weathering. Eventually, however, they realized that these holes, often in grains of quartz (a relatively hard mineral), were drilled by fungi, a process that science writer Jennifer Frazer describes as "fungal mining operations."[20] Remarkably, such operations have been happening around the clock for more than a billion years, almost everywhere around the planet including deep on the ocean floor. Even when one considers the present day, when humans are radically transforming the world by mining for hundreds of substances, fungal mining activities dwarf those of humans. Fungi do not create the same kind of environmental problems associated with human mining: the open pits, leach fields, or catastrophic spills. Fungi's mining activities are largely ecologically positive, as they provide access to an incredible amount of minerals for plants and the organisms that eat them. Although it is hard to compare quite different positive and negative environmental effects, in terms of net planetary effect, fungal mining could be more important than human mining. I suggest that the main reason why this is just emerging as a new scientific discovery is the tendency to imagine fungi as relatively passive, not as powerful world-makers.

At first, I assumed that one reason fungi's critical roles have been underappreciated in many accounts of biology was because their ways of living were invisible and inaudible to humans. But biology textbooks focus almost exclusively on only one of fungi's roles—as "rotters," who break up dead plants and animals. Rotting is another fungal activity that is portrayed as relatively passive: "rot happens," but while it may be silent to us, such rotting places may be buzzing with sounds and activity. I suggest, however, that rotting is in fact another highly active form of eating, which takes as much effort and engagement as eating minerals. Fungi, like other organisms, likely possess varying amounts of skill in acquiring food, and they learn through practice. While we are used to describing humans as dynamic actors whose abilities are shaped by their own experiences, we often imagine that other species have fixed responses that are neither learned nor enskilled, which diminishes a sense of their liveliness. Imagining fungi as active actors is even harder than imagining animals so because we don't see fungi acting in animal-like ways. Yet their world-making liveliness in such behaviors as mining and rotting are highly active processes that are utterly essential to life; in the process of their own eating, they unlock nutrients and bring them from sinks into cycles for themselves and for others.

Making Land Safe for Plants

After nearly half a billion years of fungi breaking down and decomposing terrestrial rock, plants finally joined fungi on land, and the result was spectacular. Current molecular genetics estimates that the joint travels of plants and fungi began at least six hundred million years ago, and with these new partners, the world of fungi expanded rapidly. Although this early date remains speculative, remarkably enough, there is a rich array of fossilized fungal connections to plant roots that date back 460 million years. Until 2019, these were the oldest-known fossilized fungi. But then a graduate student in Brussels made a startling discovery: small, almost undoubtedly fungal bodies embedded in a rock in Canada near the Arctic Circle that could date to a billion years ago, doubling their known existence and revealing an incredible time gap to explore.[21] The organism possessed structures associated with living on land, but researchers are not sure if it lived on land or in water. Of course, it takes a rare conjuncture of many factors for a fungus to fossilize, as their structures are extremely delicate and perishable. Scientists have only recently begun to look for these fossilized fungi, in part because previously they did not imagine that fungi could fossilize.[22] Looking for fossilized fungi is similar, in fact, to hunting for mushrooms: both fossil and mushroom hunters tend to find many more when they have a clear "search image" in their minds, when they know what to look for.

Plants' migration onto land brought with it numerous challenges. In the ocean, fungi and plant bodies were constantly surrounded by water, and drying out was not a concern. On land, however, early plants could live only in places with constant moisture, and much like some mosses that can still be found today, they absorbed water and nutrients through their skin. The first land plants were challenged by exposure to the desiccating powers of sun and wind and thin soils unable to retain much moisture between periods of rainfall.

Over time, as fungi continued to break down the rocky surface and create soil, land plants developed roots. These first roots were relatively coarse, more effective at anchoring plants in soil and holding them upright against the wind than in acquiring water and nutrients.[23] The particular partnership that fungi have developed with plants was to connect fungi's fine hyphal threads with the plant's coarse roots. This partnership is called mycorrhizae (for "fungus" and "root"), which I explain in more detail in chapter 2. Such partnerships are encounters with major consequences, where each partner is fundamentally transformed through the engagement. They meld together, joining their circulations. Fungi and plants created two main forms of mycorrhizae, arbuscular mycorrhizae ("tree-like") formations (AM) and ectomycorrhizal mycorrhizae ("outside-mycorrhizae") formations (EM). This long history is now embedded in each organisms' DNA, and the potentiality of their relationships shapes the body of each. For example, pine trees create short roots whose purpose is to house the EM fungi (fig. 1.2). In other words, the shape of trees' bodies, in this case the roots, is shaped by anticipation of their fungal partner.

Fungi are, in short, a critical conduit between plants and the underground world. Fungi dramatically expand the surface of a plant's roots through their often microscopic hyphae. These can acquire much more water than plants, especially during critical periods when the soil is relatively dry and the remaining moisture is tightly bound to the soil matrix through forces of attraction, so more force is needed to access it. As I will explain in more detail later, these hyphae can also obtain minerals that plants are not able to extract from the soil. In turn, plants send up to a quarter of their photosynthetically fixed, carbon-based sugars downward to fungi that cannot make sugars on their own.[24] In other words, when you look at a tree, one out of every four leaves is devoted to maintaining fungal relations.

FIGURE 1.2. A view of fungi in the ecosystem. From left to right: An oyster mushroom living on a log, capturing a nematode. A pine tree attached to porcini mushrooms by a mycelial network. Indian pipe, a nonphotosynthetic plant, "hacking" this network. An oak tree also connected to sporulating matsutake mushrooms as a sciarid fly lays eggs in their body. Like Indian pipe, a candy cane plant tapping nutrients from underground mycelium and drinking only from matsutake. Image by Saki Murotani.

Solving the Puzzle of Lignin and Ending the Age of Fossil Fuels

Returning to our fungal history of Earth, hundreds of millions of years ago as plants and fungi lived in partnership, they expanded rapidly. For a long time, fungi were mainly plant enablers, fostering plant growth through mycorrhizal relations and also decomposing their herbaceous bodies when they died, literally preparing the ground for the next generation. Eventually, plants invented a new complex molecule that allowed for tremendous upward growth: lignin. This complex and tough structure became one of the key building blocks of wood. Without any kind of vein-like tubes or woody material, plants can reach only a very limited size, about the height of a tulip. The development of lignin completely transformed plant bodies into towering structures that could capture copious amounts of sunlight, which in turn created a whole new scale of vegetative potential: the tree. Some became massive, and in places most conducive to tree growth, they could live for centuries, or under ideal conditions, potentially forever.[25]

Once lignin was developed, massive forests arose virtually overnight, where previously, for many millions of years, there had only been short plants. Previous to lignin, almost all organic matter had become part of a flow of nutrients through the ecosystem. But after the arrival of lignin, for hundreds of millions of years nothing could break it down. This caused trees to become a new phenomenon: a biological dead end. That is, even trees that died from fire or disease could not be incorporated back into the cycle of life. Even when trees died, they were structurally sound and took a long time to fall. When they fell, they could remain hard for decades, if not centuries. Over time, these fallen logs built up and up, until, in some places, it is likely that there was no room for new trees to grow. After repeated floods buried these swampscapes, massive amounts of undigested plant matter piled up and became subject to great forces of compression. In places with perfect conditions, the trees with their lignin and cellulose were compressed and then eventually turned into coal and petroleum.[26]

Previously, the dominant story was that the carbon age ended mainly because of a shift in climate. There are now several contenders, and one theory that is gathering some prominent supporters is that fossil-fuel-making was brought to a halt when fungi remade the world by learning how to digest lignin; this new ability set off a feeding frenzy in which massive piles of dead wood around the world were consumed and transformed into simpler forms of organic matter. Thus, the amount of coal and gas underground is proportional to the time gap between when plants developed lignin, which allowed them to become trees, and when fungi learned how to disassemble lignin and turn it into food.[27] If fungi had immediately figured out how to break lignin down, there would have been no gas or coal. Humans could still burn wood, but the whole fossil-fuel-powered world would never have existed; likewise, we would not now be faced with climate change. Every time you use a machine that burns gas—in order to fly or drive, cook your meal, or heat your home—think of the fungi that both started and stopped the creation of fossil fuels. In turn, some of the same digestive practices that fungi developed to break down lignin also enabled them to eat petroleum, including in oil spills, which I discuss in more detail in the conclusion.

How Fungi Made Soils Rich in Organic Matter and Minerals

After fungi learned how to break down lignin and cellulose, they actively began to enrich the planet's soils by breaking down the very same plant tissues that they had played a major role in producing. This, then, was the third major evolutionary step in the conditions that fostered land plants. To recap: first, fungi spent perhaps six hundred million years converting rock into minerals and creating the basis for soil before plants arrived on land. Second, on land, fungi formed intimate relations with plant roots, which greatly enabled plants' ability to grow in a much wider range of climates and conditions, while simultaneously providing fungi with food. Third, fungi learned to rot wood, a substance otherwise unpalatable to almost any other organism, and turned wood into organic matter and humus that created rich soil. This soil, in turn, became the basis for a new niche that spread virtually everywhere around the world. These evolutionary dynamics created life not only for plants but also for diverse forms of animal and bacterial life, which in turn changed the soil through their own activities.[28]

Periodically, over the deep history of Earth, such complex relationships were radically simplified during major extinction events, when many plants and animals died off. Around sixty-six million years ago, a massive asteroid struck Earth near Mexico, creating a 180-kilometer-wide crater; the debris from the collision clouded the entire planet, likely plunging the planet into a two-year period of cold and darkness. Under these harsh conditions, almost no animals weighing more than fifty-five kilograms survived.[29] Until the year 2018, I had mainly read about this mass extinction in terms of its impact on animals, with little attention given to plants and none to fungi. New accounts, however, suggest that in addition to the death of big animals, many plant species starved from lack of sunlight. This mass plant death, in turn, created a flush of fungal life. In these devastating conditions, saprobic fungi were some of the few organisms that flourished because they feed on dead plants and don't require sunlight. Scientists have found a layer of fossilized fungal threads and spores in the sedimentary rock where the remains of conifers and flowering plants were broken down by massive numbers of fungi. Although these saprobic fungi could flourish, the fungi that were mycorrhizal, which need living plants to survive, likely suffered greatly with the death of their green allies. Unlike large dinosaurs, however, mycorrhizal fungi did not die off completely. If they had gone the way of the dinosaurs after the asteroid, the current world would be completely different. After such a dramatic time of extinction, the continued survival of mycorrhizal fungi remained a major force in the proliferation of plant life and the diversification of many forms of life that depend on these relations.

Fungal Diets

The decline of the dinosaurs that this asteroid initiated led to the rise and diversification of mammals, who expanded into new niches as their bodies evolved, in part in reaction to fungi. Unlike fish, amphibians, and reptiles, which are (to borrow a concept from the ancient Greeks) "cold-blooded," mammals represented a bold new experiment in bodily design: they were "warm-blooded." Many theories have been proposed for why this warm-blooded design emerged and survived, especially considering the major disadvantage of requiring massive amounts of food (as much as six to eight bites out of every ten bites) even while still, just in order to maintain a constant level of high body heat. Like all theories of evolutionary biology, these are speculative. Scientists cannot create experiments to definitively know why things are the way they are today, why certain paths were created many millions of years ago. Yet some experiments can be carried out.

One hypothesis is that warm-bloodedness evolved as a defense against fungi, as many fungi find our bodies too warm for comfort. This hypothesis was recently supported by scientists working with bats on the devastating case of fungal white-nose syndrome. Scientists realized that bats are infected only while they are hibernating in cold caves, where their body temperature drops. If bats that would otherwise succumb to the fungus are awakened and exposed to warm temperatures, they can fully recover from the fungus,[30] demonstrating the role of heat and an active immune system in fighting fungal disease. Other laboratory experiments tend to confirm that the fungi most threatening to us cannot thrive within our bodies, as we are too warm.[31] Indeed, fungi that might have been shunted away from warm mammal bodies may have been a lethal killer of cold-blooded reptiles, such as dinosaurs. We see this happening today, in the case of fungi's devastating impact on amphibians, such as many frog populations.[32] Now, as then, fungi are busy forming a wide range of relations with other organisms.

At some unknown points in evolutionary time, fungi had intimate relations both with plant roots and within plants' aboveground bodies.[33] Fungi that live within leaves, part of a world within plants, are called endophytic fungi ("endo" for middle and "phytic" for plant), a term first coined in 1866. It is likely that many of the important chemicals that plants have—which help them live in extreme temperatures and deal with salinity and insect predators—are created by endophytic fungi and bacteria. It turns out that endophytic fungi are plentiful; since they were discovered, this study has exploded, and scientists have found endophytic fungi within every plant sample they have tested.[34] Historically, most research on leaf-born fungi focused on visible species that harmed plants, and it was first suspected that these newfound fungi were also harmful.[35] Now, however, researchers see fungal-plant relationships as diverse and complex, with most of these helping plants flourish.

Many of these fungi may play a major role in how the plant actively protects itself against insects, bacteria, and other fungi; fungi are a major component of the plant microbiome.[36] When plants are being eaten by bacteria or insects, some endophytic fungi can quickly produce chemicals to repel or even digest these potential predators. Thus, fungi have provided plants with antibacterial capacities for likely millions of years. For humans, we have been using fungi medicinally for thousands of years, but it was only in 1928 that we intentionally used fungi's bacteria-fighting powers to create the basis for the most important medical discovery of a century: antibiotics.[37] Scientists now recognize that fungi's ability to create toxins with unique alchemical capacities is unparalleled. Fungi have long excelled in finding ways to turn nearly everything into food, sometimes allowing erstwhile prey like plants to turn the tables and attack or fight back against potential predators.

Fungi are also found within animal bodies, playing little-studied roles. Whereas scientists had typically perceived most animals as individuals, there is growing recognition that all animals harbor teeming worlds of life within them, including bacteria but also fungi and others, so that each of us is a conglomerate of species. Animals are especially reliant on microbes to eat, cows being the most famous example. Many insects rely on forms of microbial life to turn food into Food; termites, for example, are completely dependent on fungi harbored in their digestive tracts, which enable them to eat wood.[38] Human bodies, as well, are thick with forms of microbial life, and these organisms are now considered essential to our digestion (a recognition seen in the recent invention of the term "probiotics"). As biologist Rob Dunn suggests in his wonderful book *Never Home Alone: From Microbes to Millipedes, Camel Crickets, and Honeybees, the Natural History of Where We Live* (2018), human bodies are shaped through their everyday encounters with a myriad of species that live within our bodies and within our homes.[39]

At the macro level as well, as I will describe in more detail in the next chapter, many fungi have strong relations with many animals, attracting them and using them as spore carriers, much as how plants use flowers and fruit to get animals to carry their pollen or seeds: fungal spores hitch rides within the animal's digestive tract or stick to their hair or feathers. As far as scientists know, although many fungi have worked out arrangements with woodpeckers, beetles, and ants to carry their spores, they haven't worked out a special relationship with humans as spore carriers to other locations, but humans support a rich mycobiome. Sometimes humans intentionally move fungi, such as when they bring button mushroom mycelium from Europe to Pennsylvania to raise on pasteurized compost, and when they transport shiitake mushroom mycelium from Japan to elsewhere to raise on oak logs.

On the other hand, people have also unintentionally transported fungi around the world, and such chance importations have played major roles in changing the course of history. Some fungi destroy the structural materials humans rely on (like wood for ships, for the support of train tracks, and for lumber in houses) or the plants that we grow for food. For example, dry rot, a kind of fungus that was accidentally transported from its original home in the Indian Himalayas to other parts of the world by British colonists, was the scourge of the British Navy and spread from ships to houses around the world. Fungi can also feast on living plants, and our methods of monoculture agriculture have fostered fungal growth by giving fungi easy access to millions of genetically similar plants, so once a fungus starts eating one of these plants, it can easily spread by spore to others nearby. One of the most well-known examples is the devastation of Ireland's potato crop. By 1600, potatoes had been brought from South America to Europe, where they thrived. Yet, like dry rot, a potato-eating fungus was unintentionally transported to Europe, where it started to wreak havoc in Ireland's large potato fields. By 1845, this dreaded rot turned vast fields of potatoes into black goo. Originally viewed as a kind of fungus, this organism is now considered "fungi-like." Another important organism still viewed as a "true fungus" is called "rice blast," which every year eats as much rice as could have fed sixty million people, the population of Italy. Such predators of our foods constitute a massive challenge for maintaining global food supplies. Legions of agricultural scientists are busy trying to breed new fungus-resistant varieties and brew new fungicides, which often are nonspecific, wiping out beneficial as well as harmful fungi and with many unknown effects.

Beyond the handful of plants that humans have turned into crops, fungi also play a major ecological role in shaping the global distribution of other plant and animal species. Fungi have learned to become parasites of almost all known taxa of plants and animals[40] and have thus coevolved in tandem with their immune and defense systems. For example, some beetles have evolved a shell that is so tough that few fungi can easily break it down—an impressive feat considering that fungi have such powerful alchemical powers.[41] Yet beetles still need to breathe, and fungal hyphae have learned how to gain access to the beetles' soft insides by entering through small breathing holes in the shell. Although mammals may have developed high temperatures in part to render themselves less vulnerable to fungal species that prefer cooler habitat, humans are still susceptible to fungal attack. Mainly we think of these as relatively trivial attacks, such as athlete's foot, ringworm, or a minor yeast infection, yet fungi are likely far more powerful than we yet understand.

Most discussions of fungal threats to human health focus on people mistakenly eating toxic mushrooms, but microscopic fungi are an important and largely invisible source of human mortality in other ways. Previously, Western medicine gave little attention to the active role that fungi played in human mortality, which parallels the broader history of myco-ignorance. These days, however, there is a growing recognition of the important roles that fungi play in human health, along with new tests to determine their presence in and on our bodies. Some scientists are now suggesting that fungi may be a more important vector of human death and disease than are more commonly recognized vectors such as insects (including malaria- or zika-bearing mosquitoes and schistosomiasis-bearing snails), bacteria, or viruses. For years, people with compromised immune systems from HIV/AIDS were dying from disease, and when antibiotics (which kill only bacteria) didn't work, doctors were stymied. Finally, a researcher suggested that the disease impacting them was fungal, and patients recovered after switching medicines from an antibiotic to an antifungal.[42] According to this new understanding, disease is caused less by the absence or presence of bad germs (bacterial, viral, or fungal) than as a result of an imbalanced ecology.

Such a transformation is starting to happen, but many of us remain caught in a pathogenic framework; and there has been a notable lack of research on how a fungal presence in our bodies might contribute to health. For example, at a conference, the microbiologist Margaret McFall-Ngai once talked to a fellow researcher who studied the *Candida* fungus in humans. She asked about its possible role in fostering health, and the candida researcher replied, "Good question; I've been working on it for thirty years but always assumed it was only a potential pathogen." This remark ended up increasing the researcher's exploration of the positive role of candida in our digestive tracts.

In this way, scientists like McFall-Ngai are pushing the study of microbes, which has historically taken a pathogenic approach, toward an ecological approach. Her pathbreaking research has shown us how bacteria played a significant if largely invisible role in the evolution of animals. Whereas many neo-Darwinian approaches to evolution examine an organism in relative isolation, McFall-Ngai argues that evolution does not work solely at the level of the genetics of individuals but in relationship to their larger microbiome.[43]

It is quite probable that fungi, like bacteria, are a part of a larger microbiome, and that the roles they play in shaping the diversity and expressions of life are as powerful as those of other significant evolutionary players.[44] Thus, we should not imagine that fungal capacities to destroy are a purely negative phenomena, because deadly fungal powers also foster the abundant diversity of life. For a long time, botanists puzzled over how to explain why tropical forests are often diverse and why temperate forests are relatively homogeneous. One part of this history was revealed in the early 1900s after the industrialist Henry Ford planted a vast plantation of rubber trees in the Brazilian Amazon to "grow" tires for his Model T car. In the wild, rubber trees grow far apart, and rubber tappers have to spend a lot of time moving between trees. To make the harvest more efficient, he planted cleared forest lands with vast numbers of rubber seedlings. Soon, however, a particular fungus destroyed all the closely growing seedlings, revealing that the widely spaced distribution of rubber trees in the natural forest was shaped by the presence of fungi. Botanists now think that similar dynamics may be true for many other tropical trees: fungi eliminate those that grow closely together, and only those that are far enough apart survive.[45] Such histories might offer a portent for human's efforts at creating monoculture plantations. As we saw with potatoes and Ford's rubber trees, we are witnessing a contemporary crisis of banana plantations as fungi consume them, jumping easily from plant to plant. Our industrial large-scale agriculture creates perfect conditions for fungi's rapid proliferation.

Conclusion

For billions of years, the newly founded planet Earth was a land of water and rock, without soils; but for half a billion years, some forms of fungi explored and interacted with the terrestrial world. By slowly secreting acids that digested the rock, fungi multiplied, diversified, and spread. Then new forms of plants, likely moss-like organisms, came out of the waters, together with fungi, and a miraculous partnership between plant roots and fungal bodies was formed. New waves of plant and fungal life spread and diversified for at least one hundred million years before plants created a stunning new invention, lignin. Then, around four hundred million years ago, lignin enabled plants to become soaring structures—trees—making new habitats and accelerating the breakdown of rocks into soil. For sixty million years, tree bodies piled up on the landscape and fell into swamps, until fungi finally discovered how to unlock lignin and eat dead wood, initiating a seismic change in how ecologies functioned. Thus, fungi helped create the planet's first forests. Fungi have kept these forests thriving today by assisting plant roots and leaves, as well as through their critical role in breaking down the dead and turning their bodies into nutrients for the living.

In the realm of animal evolution, I showed how some scientists argue that warm-blooded animals like mammals and birds might have evolved in part as a way to ward off fungal growth. Evolution is often coevolution, always in relationship to other species that we eat and that eat us. Ironically, as humans are warming the planet and the ambient temperature grows, fungi are adapting. Thus, the gap between the air temperature and our warm-blooded bodies is shrinking, making it easier for fungi to survive within our warmth, including drug-resistant forms of *Candida auris*, a deadly species of the candida fungus that lives in many human bodies, often without harm, unless it moves into the bloodstream. As we have seen, fungi have affected more than just animals, and they have been quietly shaping the planetary evolution of animals, plants, and others for hundreds of millions of years. I believe that insights into these dynamics, which are increasingly understood by biologists, are likely to continue to grow as more people start to comprehend fungi as lively actors.

The following chapter dives deep into the hidden realm of fungal world-making at a smaller scale, looking at a range of specific kinds of mushrooms and showing how they engage the world, form intimate relationships with plants, and are hunters as well as prey. The chapter explores the lifeworlds of some mushrooms, in what I call their "every-day fungal world-making." Their interpretation of the world around them shapes how they engage with a panoply of species, making worlds both with and without humans.

CHAPTER TWO

Everyday Fungal World-Making

They found themselves, like any first creatures, lost.

Without means, they were unable to survive by anything other than what was in the immediate surroundings. They ate what grew. They planted nothing. They never left home. There were many dire moments, until they found the animals. The first time would have been accidental. A young one caught an animal and rode it out somewhere, the way a storybook character might ride a boat down the river and out to sea.

With time though, they learned more tricks. They waited where the animals came to feed. They found them where they slept. Soon they were riding them all the time, clinging to their dark bodies as they darted here and there into the unknown. Good luck took them to more food. Bad luck killed them. Time, birth and death made good luck more common.

Over years, they reined their new beasts in until, as is the case today, the steeds go out and gather food and bring it back. The fungi grow and wait. They have become fat kings whose success can be measured by the number of their beasts. And they are not few. These protagonists, each one a fungal herder, have evolved multiple times. They are exotic, and yet in some contexts, far closer to home than you might believe.

**—Rob Dunn, "Five Kinds of Fungus Discovered to Be
Capable of Farming Animals!," 2012**

Over hundreds of millions of years, as I outlined in the previous chapter, fungi have changed the course of planetary evolution. Yet even events as powerful as these would scarcely register as "agency" in the human-centered terms with which we typically imagine agency: that is, as the intentional actions of *an individual*. These major transformations were not singular, discrete events (like the massive meteorite that likely spelled the end of the dinosaurs), but a series of billions of actions over a long period of time, which, as individual acts, would be largely invisible in terms of the conventional human-based conception of agency. However, these acts can and do accumulate, becoming what some call "cumulative agency."[1] Thus, while I began to recognize that fungi were indeed "world-*changing*," I still wondered if it was possible to understand their actions as "world-*making*." I was especially interested in seeing how fungi engaged the world in dynamic ways that expressed their liveliness. Put somewhat differently, what would a matsutake's life story look like if it were written in the active voice? I envisioned telling the story of its life cycle, from its inception as a spore until it made spores itself. I wanted to show its lively path through life, from being carried by the wind, to making relations with trees, to creating a fruiting body, to being eaten by small soil-dwelling insects, to again growing into a mushroom. But I could not find enough written about the matsutake life cycle to build a complete story; I could find only glimpses into short moments in its life, as described by the few researchers who occasionally wrote about it. I was delighted to find the work of biologist Rob Dunn, whose account of fungal life serves as the epigraph to this chapter, because he flips the usual script of "insects that farm fungi," and instead explores the potential agency of the fungi. This motivated me to expand my quest, drawing examples from across the wildly diverse kingdom of fungi, as I did in the previous chapter. This chapter, then, delves into a range of world-making actions that various fungi carry out (while I save the specific details of matsutake world-making for chapter 4).

Before jumping into this chapter, I'd like to offer a brief road map of my exploration into these issues. Initially, my goal was to explore scientific writing on fungi and, specifically, to search for forms of liveliness and agency. After coming up short, I reversed my approach in an effort to understand why I was mainly finding descriptions of fungal passivity—of fungi "reacting to stimuli" rather than making worlds. This led me to consider how subtle (and not so subtle) aspects of the English language itself worked to reinforce a Great Divide between humans and all other forms of life. Indeed, I realized that long before biology as a discipline came into being, linguistic conventions had already succeeded in effectively de-animating the nonhuman world. Next, in trying to see why there might be a particular bias against fungi within biology, my attention turned to thinking about how fungi were understood in England, the country that has arguably been the most influential in shaping the version of biology that remains dominant in the world today. I looked for connections between everyday British understandings of fungi and how fungi are represented in scientific textbooks. As I explored descriptions of fungal activities, such as movement, eating, and behavior, I realized that biologists typically used animal-defined understandings of these activities, which left out plants and fungi. Searching for glimmers of a more lively approach, I found one example in "niche construction theory," which offers a way of seeing the world that emphasizes the active agency of organisms in shaping their immediate environments. Although this approach has, to date, mostly been applied to animals, I saw that it offered possibilities for taking a lively, world-making approach to fungi.

The Grammar of Animacy: The Role of Language in Perpetuating the Great Divide

Endeavoring to find lively ways to describe fungal lives, I searched widely throughout the mycological literature. Occasionally I found skilled science journalists like Jennifer Frazer and Ed Yong, who could transform scientific articles into engaging essays that could be understood by a lay audience; but I discovered that scientists often cringed at how nonscientists "anthropomorphized" nonhumans by attributing human characteristics to them.[2] This led me to wonder: What are the main criteria that determine what counts as science?

Eventually, I came across the work of Potawatomi botanist Robin Wall Kimmerer. She reflects on how her own people's language emphasized the active presence of nonhumans in the world, including fungi:

Puhpowee . . . translates as "the force which causes mushrooms to push up from the earth overnight." As a biologist, I was stunned that such a word existed. In all its technical vocabulary, Western science has no such term, no words to hold this mystery. You'd think that biologists, of all people, would have words for life. But in scientific language our terminology is used to define the boundaries of our knowing. What lies beyond our grasp remains unnamed.[3]

When I asked several mycologists about the term *puhpowee*, they told me they appreciated its specificity and onomatopoeic qualities: the last syllable *wee* sounds like something that is moving, expanding up from the earth and into the air. The closest scientific term they could think of to describe a mushroom's growth was *hydraulics*, whereby a mushroom uses water pressure to expand its parts. Others thought of "turgor pressure" to describe the pressure within an organism's cells, but this did not necessarily capture all the dynamic forms of growth. Each of these terms tends to reduce this action in terms of engineering or physics. Perhaps the lay term *mushroomed* in some way describes this rapid and active growth, by turning a noun into a verb, but in the past, this term has often been used to describe negative phenomena such as taxes or crime, and the term "mushroom cloud" superseded the old name for the horror of a nuclear blast: "cauliflower cloud." Kimmerer's term *puhpowee*, however, resonated with a broader pattern found in other Indigenous languages that describe many organisms as actively *encountering* the world.

In fact, it was Kimmerer, alongside other Indigenous scholars such as Kim TallBear (Sisseton Wahpeton Oyate) and Zoe Todd (Métis), who helped me reflect back on the particular consequences of the potential relationship between worldviews embedded in European languages more generally and within scientific language specifically.[4] In particular, Kimmerer helped me to read the scientific literature in a new light and to notice the ways that scientific English typically works to strip away the animacy of animals, plants, and fungi—not only for scientists but also for nonscientists, who use distancing language such as the pronoun "it" or "that" for nonhuman animals, which effectively casts them as objects rather than subjects. As you have likely noticed in this book, I use pronouns of personhood for our kin.

I'm not necessarily calling for an end to objectification but trying to point out and question what we might describe as the often invisible everyday practices of human exceptionalism and human supremacy that are created through language and that reinforce conceptual categories. Let's briefly consider gendered pronouns. In the past decade, there has been an expansion in thinking about gender; for example, there is now an accepted third category for humans, *they*, which some now use as a default when they are not sure how another individual identifies, or as a self-described term. We still, however, draw the line in another binary between humans who deserve gender pronouns and nonhumans who don't. Even when we "know the sex" of a wild bird, for example, people will usually avoid the term "she" or "he" and refer to the bird as "it." One of the few exceptions is animals who have been brought within the family circle, such as dogs, cats, and other pets. Indeed, the way language reinforces a sense of a Great Divide at all levels is even deeper than the differences in gender between he/she/they and it, as evident in the custom of referring to persons as "who" and nonpersons as "that," a convention that I am eschewing in this book.

One of the deeper underlying problems with mainstream biology is that it relentlessly describes the nonhuman world in passive and mechanical terms, as not fully alive and dynamic. As Kimmerer reveals, this conceptual framework and attitude toward other organisms exerts a powerful effect on how we construct biological knowledge and treat other organisms as things, or as resources solely for human experiment, use, or profit.

Although I have often heard people describe scientific articles as "dry," I think they might more accurately be described as "mechanistic." Almost none of the scientific readings I came across engendered a sense that fungi actively engage the world; instead, they are described in a mechanical mode of stimulus-response reactions—stimulated, for example, through chemical triggers. This mechanist notion of reaction and trigger is one of the main ways in which the current scientific orthodoxy renders organisms as passive, making it hard to see them as active creators and participants in the world—that is, as world-makers. How, then, might we begin to see fungi as dynamic and engaged organisms?

With this question in mind, I set out to find alternative frameworks within the scientific literature, along with explanations for the rise and dominance of mechanist thinking. As I describe here and in the following chapter, in order to learn specifically how mushrooms engage the world, I turned to the work of scientists who spent their lives watching and experimenting with plants and fungi. If we want to understand fungi as world-makers—as carrying out daily activities, and behaving and interacting with other species—what might Western biologists tell us? Interestingly, my openness to the idea that fungi might interact with other species had begun years earlier while working in China with ethnic Yi mushroom hunters in Yunnan Province—another group of people who spend their lives observing and experimenting with fungi. I was curious if I could find a similar approach among Western scientists. In looking for such similarities, I also drew on my training in the anthropology of science to look critically at the production of Western biological knowledge, in order to ask: How do the metaphors and perspectives that dominate modern biology shape what we know, or think we know, about the world?

Challenges within Current Scientific Orthodoxy

When I was exploring the scientific literature, eagerly hoping to find lively fungi, I was reminded of a group activity I had engaged in many times with other mycophiles. This activity is called a "foray," wherein members of a mushroom club try to find as many species as possible, whether edible or not, to see which species of fungi are fruiting in a certain time and place. It's a kind of treasure hunt, an exploration into the unknown, and participants must have the capacity and readiness to be surprised. Yet in my forays into the scientific literature, reading many specialized mycological books and articles and numerous introductory biology textbooks, I had a hard time discovering any good examples of lively fungi. Why was this so?

There were a number of challenges, and here I explore two. The first is specific to fungi and relates to how they have been relatively ignored, stigmatized, and feared in England, a country that has had, since Darwin's rise in the late nineteenth century, a disproportionately powerful influence in shaping contemporary theories of evolution and science and scientific writing more generally. Might this culturally specific attitude play a role? The second is that the biological sciences are structured by a mechanistic framework that tends to work against my interests in understanding the liveliness of fungi. I did, however, find a handful of exceptions to this trend, and I highlight a few examples below.

The British Legacy of Mycophobia

Within the world of mycology, there is a now famous story about an Anglo-American investment banker and a Russian doctor taking a walk in the woods. This event, which sounds like the beginning of a classic joke, turned into a series of events that helped precipitate the Western fascination with psychedelic mushrooms and expand interest in fungal powers. It also revealed the diversity of cultural attitudes toward fungi and a particularly strong and specifically British antagonism toward our fellow opisthokonts.

In 1927, Gordon Wasson noticed that his fiancée, Dr. Valentina Guercken, rushed from the path to exclaim over a cluster of mushrooms that seemed like some beloved edibles from Russia. Wasson, however, was fearful and refused to eat them. They were puzzled by this stark difference in their attitudes. This event launched their joint lifelong quest to better understand the diversity of human relations with mushrooms.[5] After several years of research, they coined the terms "mycophobic" and "mycophilic" to describe two distinct cultural attitudes: mushroom fearing and mushroom loving. They were surprised to note that while many cultures loved mushrooms for diverse reasons, Britain stood far on the mycophobic side; and indeed, many assert that the British are the most mycophobic of all the world's known cultures.[6]

Since learning about this widespread negative attitude of the British, I have wondered how such mycophobia might have influenced scientific understandings of the role of fungi in the field of ecology. Further, does such a sensibility continue to haunt biological studies, and has it also become embedded in the English language?[7] The possible influence on English is especially important because it is the world's dominant scientific language, with 80 percent of the world's scientific findings published in English, as are all fifty of the top-ranked scientific journals.[8] As well, I found that a number of influential biology textbooks in other languages were also translations from English.

Researchers in British settler colonies, such as Canada, the United States, Australia, and New Zealand, confirm that widespread mycophobia remains strong among lay people, which certainly resonates with what I know from my own upbringing. I grew up in an Anglo-American context where few knew much about mushrooms, and many felt uneasy about them, certainly too nervous to try eating wild mushrooms at a restaurant, let alone ones they had found along a trail. Still, the origins of British mycophobia are perplexing, given that neighboring countries such as France, Italy, and Russia show a great love for mushrooms and mushroom hunting.

The Wassons' depiction of British mycophobia was not a surprise within the UK itself, where fungi have often been associated with death and rot and referred to as the "pariahs of the plant world."[9] Going back to the 1850s, Miles Berkeley[10] was an important explorer of fungal lives and described his fellow citizens' negative attitude toward fungi this way: "From the poisonous qualities, the evanescent nature, and the loathsome mass of putrescence presented in decay of many species, [fungi] have become a byword among the vulgar and are frequently regarded as fit only to be trodden under foot."[11] Thirty years later, William Delisle Hay, a fellow of the Royal Geographical Society and student of mushrooms, reconfirmed Berkeley's writings about British attitudes:

> All mushrooms . . . are lumped together in one sweeping condemnation. They are looked upon as vegetable vermin only made to be destroyed. No English eye can see their beauties, their office is unknown, their varieties not regarded. They are hardly allowed a place among nature's lawful children, but are considered something abnormal, worthless and inexplicable.[12]

When British mushrooms were not simply lumped together, they were divided into two categories: "mushrooms" and "toadstools." The former term described the edible variety and the latter described poisonous mushrooms, indexing their strange nature and affiliations with another poisonous organism (toads) that had long been associated with witchcraft. In that same essay, Hay coined the term "fungophobia" to describe the fear of toadstools.

This existing fungophobia was reinforced after it was realized that microscopic fungi created powerful diseases that killed important crops. Fittingly, as if to reconfirm and extend the existing British mycophobia, the first major breakthrough in discovering fungal diseases took place in the UK. After many years of British colonial expansion into Ireland, much of the population subsisted almost entirely on potatoes. The potato blight, from 1845 to 1846, caused at least one million Irish people to starve and two million to emigrate. Even today, Ireland's population has not recovered. Eventually, the disease was attributed to a water mold that scientists saw as fungi. This relationship, discovered by Miles Berkeley and others, was one of the earliest wake-up calls for the power of fungi or fungi-like organisms to quickly kill off a large number of plants.

Soon, many other fungi were found to harm plants that people grow for food and timber. This led to a situation where, by the twentieth century, European forestry and agriculture textbooks mentioned fungi in almost exclusively negative terms; the vast majority of funded mycological research in these fields was based on fungi's potential for disease, not potential benefits.[13]

This legacy of mycological study, far more influenced by a pathogenic model than an ecological one, has significantly shaped the wider realm of mycological knowledge. I became aware of this embedded bias only after a conversation with pathbreaking microbiologist Margaret McFall-Ngai, who described how a similar model had shaped the field of microbiology. In her estimation, at least 80 percent of microbiologists focus on disease. She also pointed out that some common scientific terms reveal this disease-based perspective, such as the term "infection," which describes *any* connection between bacteria and plants or animals, whether harmful, helpful, or neutral. Fungal connections are also described as infections; when I mentioned to one mycologist that infection might convey a negative connotation, I was told it was neutral. Yet infections are usually defined as a "negative interaction" that causes harm.[14] The term presupposes that fungi are harmful, even when such connections are made between a mycorrhizal fungus and a tree root, from which both benefit.

The situation could have been otherwise: if the helpful, rather than harmful, relations between plants and fungi had been emphasized, scientific research might have followed other paths. For example, more research could have been undertaken into how fungi help crops and trees to grow. Scientific knowledge of these beneficial relations was first proposed by the German botanist and mycologist Albert Bernhard Frank in 1885. He coined the term "mycorrhizae" (for "fungus" and "root"), suggesting that this was an intimate and mutually beneficial connection between fungi and plants. At the time he made these pronouncements, only twenty years after the devastating potato famine, many scientists assumed that fungi had a negative pathogenic relationship with plants, and Frank's ideas were largely ridiculed. This was also twenty years after Herbert Spencer coined the term "survival of the fittest"; and with the Darwinian notion of competition and the struggle for survival, it was easier to imagine interspecies relations based on antagonism than on mutualism.

Frank's discovery was an accidental one that occurred during his research (on behalf of the king of Prussia, Wilhelm I) on cultivating truffles. As with most attempts to grow mycorrhizal fungi, the project failed, but in the course of Frank's work, he was the first to document a relationship that many people had noticed before, but had not understood. It would take decades before many of his claims—what one of the world's most prominent truffle researchers, James Trappe, called a "revolution in thinking"—were well supported and accepted within the scientific community.[15] In the realm of agriculture and forestry, we would have to wait more than a century before significant research was carried out in the West into how one might use fungi in a widespread and positive way,[16] especially in promoting plant growth through adding or managing mycorrhizae. Even though it was not intensively studied, anecdotally many colonial officers charged with introducing tree plantations to new countries learned to enhance growth by importing soil from the tree's origin site, which likely introduced beneficial fungi.[17]

At first these mutually beneficial relations were thought to exist in only a relatively small number of plants, but by 1977, James Trappe proposed that about 95 percent of plants have an underground fungal connection,[18] a figure that remains well accepted.[19] If these relations are so widespread and important, why are they not central to our basic understandings of the world? Trappe thinks that even with some recognition, mycorrhizal studies are still relatively underappreciated within the field of biology as a whole. In reflecting on this issue, Trappe supports Frank Ryan's argument that "Darwinism, with its emphasis on competitive struggle, thrived, [but] symbiosis, its cooperative alter ego, languished in the shadows, derided or dismissed as a novelty."[20] This might be an interesting example of where the dominance of competition as a model of biology combined with the lingering effects of Anglo mycophobia in ways that potentially impacted scientific orientations.

One can ask, what might have been different had the English not played such a substantial role in the orientations of modern science? What might have been otherwise if the main language for scientific publications had been Russian, Japanese, or Chinese—all historically and currently mycophilic places that seem to possess little mycophobia? As I describe later in the book, Japanese scientists are often working with a different model of how fungi live, in which they promote the idea of fungi as human's "critical life partners." I extend this notion beyond a human orientation to think about fungi writ large, that is, how fungi are likely critical life partners to every kingdom of life. Indeed, when mycologists talk about fungi, they typically describe their tree partner as a "host tree," which represents the tree as *the* primary agent—providing the space and the food—whereas the fungi are the less important guests. But, in fact, to flourish, both fungi and trees contribute critical materials, and fungi likely confer many additional benefits other than just food and water to trees, including disease resistance.

I also found that a number of Japanese scientists had a different approach to fungi that emphasized more agency on the part of fungi than I saw in many Anglophone accounts. One particular aspect of their approach was that the Japanese paid more attention to the role of fungi in attracting, rather than defending against, other organisms. In Japan, experiments seemed less committed to a presumption of antagonistic relationships between species. This idea finds resonances in a paper published by my research collaborators, Anna Tsing and Shiho Satsuka, that describes how Japanese scientists have more often understood humans as part of the ecological landscape;[21] Western scientific studies, in contrast, often assume that humans are intruders on nature and that all human activities are antagonistic forms of "interference." Such contrasts are not always so absolute, however. Learning from Tsing and Satsuka helped me to pay attention to a plurality of scientific approaches that were never described by scientists themselves and yet powerfully, and often subtly, informed their work. In what follows, rather than focus on a division between Western and Japanese approaches, I want to highlight a group of Western scientists who constitute a "minor tradition" within biology, which is usually committed to a mechanistic approach that understands nonhumans as relatively passive organisms.

Mechanism: Passive Organisms Living in Active Landscapes

One of the main challenges in searching for acknowledgment of lively fungi is that most of the research is conducted within a mechanistic paradigm, which makes it harder to see things from a world-making perspective. As philosopher Matthew Chrulew describes, by the mid-twentieth century, Western biological thinking had taken a "mechanistic turn," which tended to view an organism's actions as automatic and predictable.[22] Even our fellow animals, typically regarded as the most active group of organisms, are often portrayed as "passive, buffeted by involuntary instinctual mechanisms."[23] Historian of science Jessica Riskin describes how, by the 1930s, a more passive interpretation of mechanism prevailed and spread beyond scientists and science to become commonplace in Western thought writ large. One of the key correlates of this position was that "a scientific explanation must not attribute will or agency to natural phenomena" (including other animals), as will and agency were seen as human-only qualities.[24]

Within the field of biology, I noticed a few exceptions to this more passive portrayal, and one of these helped me to understand fungal practices in a more active light: "active niche theory," or "niche construction theory," is currently neglected within the field of biology, but it is a theory that may become increasingly important over time.[25] I was introduced to this theory by physiological ecologist Scott Turner in his book *The Extended Organism*.[26] In it, he argues that scientists almost always assume that organisms adapt to their environment; relatively few consider how environments are also shaped by animals' actions or, in other words, how animals perform a kind of world-making. Turner's book shows, for example, how termites create structures that skillfully use wind currents for heating, cooling, and removing potentially harmful gases. These structures engage with the local environment and shape it to the termites' benefit. In biological terms, this concept of "active niche construction" is very different from the more typical and passive view that species insert themselves into a preexisting ecological niche (such as a tree cavity or a shallow pond). Active niche creation explores how organisms actively carry out efforts to enhance existing possibilities, or produce these niches.

In turn, this perspective better reveals how organisms' actions bring dynamic landscapes into being as animals' cumulative actions may shape large-scale environments. This approach animates a landscape typically seen as fixed and enables us, instead, to view organisms as actively shaping their worlds, rather than as living a passive existence. We also see how niches are not isolated "bubbles" inhabited by just one species but can extend into large landscapes that contain many species. For example, earthworms' tunneling modifies vast areas of soil, infusing it with oxygen and changing how rainwater moves through the soil. Coral polyps build reefs, creating habitats and food sources for a wide array of sea creatures. The collective actions of worms and polyps transform vast amounts of the earth's surface, in turn creating hospitable ecosystems for many thousands of species. This contrasts sharply with the mainstream mechanistic view that often privileges an externalist account, a one-way path of causality of an active environment that directs evolution.[27] A niche construction view, on the other hand, is dialectical, a two-way path whereby the environment shapes an organism's actions while an organism's actions also reshape the organism's environment.[28] This process is, therefore, closer to a world-making view: where animals are not passive adapters to an active landscape but active organisms who are *part* of a landscape that is shaped through their presence and daily activities.

Compared to an active niche model, a world-making model not only is interested in how organisms shape their environment but also foregrounds their everyday encounters with each other. An active niche model focuses on the subset of organisms' actions that leave some kind of physical trace. A world-making perspective is broader, and while it shares a view of organisms as actively shaping their environment, it also emphasizes their creative and active relationships with other organisms, in ways that may or may not leave a physical trace. It suggests that all organisms, from bacteria to whales, live their lives and perpetuate themselves over the generations through their active engagements with others—not just as predator or prey—but through a wide range of relationships far more complex than the often reductive models of "beneficial species" or "antagonistic species."

Turner's accounts of the extended organism and the "active niche construction" theory, while intriguing, are focused entirely on animals, leading me to question the larger tendency in biology to consider only animals as active subjects. Are researchers also extending this kind of thinking to the lives of plants and fungi?[29] I found a handful of papers that looked at plants from a niche construction perspective (mainly with respect to how forests are a tree-created niche) but almost nothing on fungi. This was a frequent pattern in my quest for a livelier approach to nonhuman life. In other words, among the three kingdoms that have some forms of life visible to humans (animals, plants, and fungi), they are aligned along the spectrum of agency, where animals are the most potentially agentive and then—after a large gap—come plants and lastly—after another large gap—fungi.

Forms of Action: Moving and Eating
among Animals, Plants, and Fungi

Let's consider two kinds of actions—movement and eating—to reflect on how biology textbooks compare these kingdoms. Mobility is often regarded as a key expression of intentionality and agency. It is commonly said that animals are "mobile," self-directing their movement over land or through water, whereas plants and fungi are "sessile"—that is, they don't move.[30] This division excludes or does not recognize other forms of mobility that I will describe below. I was surprised to see that the same botanists who often advocate for increased respect for, interest in, and knowledge about plants and their abilities also seem to reaffirm this division. For example, in *Approaches to Plant Evolutionary Ecology*, the author states that "plants cannot move to escape competitive stress."[31] I expected that botanists would champion the objects of their research, but in fact they often diminish the abilities of plants; such tendencies make it harder to recognize forms of active plant movement and liveliness.

As well, mycologists almost always describe fungi as sessile organisms. Fungal movement, occurring as it does largely underground, is harder for humans to perceive as dynamic, even though, ironically, one of the qualities of mushrooms that some find unnerving is that they seem to just pop up overnight, as though instantly fully formed, whereas plants are noticed slowly growing from seed to seedling to mature plant, which for a large tree could take centuries. Likewise, even botanists tend to avoid the term "moving" to describe plants; instead, they refer to this as "growing," a term often understood as relatively passive.[32] In this light, fungi don't even seem to grow, they simply appear. Their disappearance is equally mysterious, seemingly instant and without leaving a trace.

Let's turn to eating—one of the key rubrics that Europeans used to classify the living world by dividing organisms according to a taxonomy of who eats whom. One of the main taxonomies used in biology textbooks divides the world into three main groups: plants as "producers," animals as "consumers," and fungi as "decomposers." For a long time, I accepted this division as basically true, but I realized that it was based on several conceptual problems. First, this framework describes plants as ecological workhorses that make the planet's foodstuffs. This view of plants as food embodies an animal-centric perspective, as it regards plant growth as simply producing animal food. Second, this framework describes fungi only as decomposers; in many textbooks they are quickly dismissed as rotters or "described as the janitors of nature," which obscures their equally important role as mycorrhizal companions to plants.[33] Such a classification likely draws on the long tradition in Anglo thinking of associating fungi with rot and decay.

This threefold division of the world (producers, consumers, and decomposers) is based on methods and relations of eating. It renders plants and fungi as passive compared to animals who actively move to obtain their food, which they then, through further action, harm or kill. Between plants and fungi, though, there are some differences: plants are often seen as more actively making their own food from sunlight, compared to a more passive notion of fungal decomposition. Decomposition in this rendering, in fact, seems simply inevitable, a process that just happens over time, without any clear agents or actions; I suggest that it should instead be understood as an alternative, and active, form of "eating." In tandem with the notion of passive decomposition, fungi are almost always described in the scientific literature as "absorbing" their food, which seems not only passive but also spontaneous, like a dead sponge that absorbs water. In the literature on fungi, I found almost no alternatives to describing how fungi gain nutrients other than the term "absorb," so I started to ponder other possibilities for understanding movement and eating from a world-making perspective.

If we look first at plants, we can see in the short term how plant branches and tendrils stretch and reach toward objects to climb up toward sunlight. A few plant scientists, such as Monica Gagliano, refer to plants as "foraging" for sunlight.[34] Below ground, their roots actively grow and move to seek water and nutrients. Thus, everyday movement and eating are intertwined. As Wright and Jones argue, "all organisms are 'consumers' in the broadest sense,"[35] as plants are also consumers of materials from sunlight, air, and soil (usually with fungal assistance). A non-animal-centric perspective, therefore, would question the dichotomous or

trichotomous division of the living world into the categories of producer, consumer, and decomposer.

Over millions of years, many plants move vast distances over space,
spreading their seeds by wind and gravity, and by clinging to animals'
fur (or lodging within their stomachs), thus expanding and contracting
their territory in the face of changing climates, animal encounters, and
disease. One can argue that such travels, however, are accomplished
without vegetal agency, shaped instead by external forces such as the
wind or animal carriers. Some argue that there is not really any activity
being carried out by the seeds themselves, thus their travels are explained by an evolutionary logic that privileges the shaping and selecting role of the environment. Even if we place agency elsewhere, it seems
that, to paraphrase Galileo, the plant yet (i.e., nevertheless) moves.
Within one lifetime, a plant's actions are more easily understood as self-
directed; the plant's entire body growth is shaped through its dynamic
response to the moving sun, to other plants, water, soil nutrients, animals,
fungi, and so forth. For terrestrial plants, this happens aboveground and
belowground, as shoots and roots explore new places in the three-
dimensional space of the air and soil.

Let's start to think about fungal movement using a mycorrhizal example. Spores' potential for mobility begins soon after they land on a
surface, when they actively engage with their own surroundings, making
a decision about whether or not to germinate. If they do so, they send
out hyphae that start looking for food to consume directly or for partners with which to share the food. If they are unsuccessful in this search
several times, they die. If they live and find sufficient plant partners, or
food and water, and are not killed by disease or predators, they can
flourish, and their hyphae expand even more, into what is called a mycelium. Over time, these hyphae keep growing and fuse with one another when they meet up (fig. 2.1).

In a second example, some kinds of fungi (often categorized as "prim-
itive") have spores that can swim. These spores are called zoospores (or
"animal spores") as they seemed to early scientists (and to me) to be
surprisingly like an animal's sperm, as these spores contain one or two
flagellums that propel the spores through watery landscapes.

FIGURE 2.1. Matsutake lifecycle. After sporulating, some spores will germinate and find a "plant host." They may also fuse with another sexually compatible hyphae. Looking inside the older fruiting body, one can see the aboveground parts are also made of hyphae that knot into this amazing structure, renewing the cycle. Image by Saki Murotani.

World-making practices are both diverse and highly specific; when it comes to these fungi with swimming spores, the bodies and behaviors of each species (their style of swimming, their priority for searching for either sex or food, the kind of food preferences they have, etc.) can have massive consequences in the world. Researchers can see that each has its own style of swimming, some in straight paths and others in looping arcs or with periodic jerking movements. Some are fast and some slow. Some are looking first for sex and follow each other's trail of water-diffused hormones. Others focus on food, like the chytrids that live in cow stomachs, allowing their mammal to digest grass.[36] Although first suspected to live only in water, chytrids are in fact found living in every ecosystem, in water, or patiently waiting, even in deserts, for water to come along so they can mate and eat. Some chytrids favor eating pine pollen, and if scientists sprinkle such pollen into a petri dish in a lab, they will arrive. Some scientists remark how this is almost miraculous, how these fungi seem to appear out of nowhere, even in a room located deep within a large building where the airflow is highly managed and purified.[37] Others have spores that swim after frogs. Chytrids are the fungi that have recently devastated amphibian populations around the world, an event that has been called "the most spectacular loss of verte-brate biodiversity due to disease in recorded history."[38]

Underground, hyphae embark on hidden journeys. I heard that a large mycelium could travel a few inches per day.[39] This was interesting, and faster than I thought. However, when I added up all the millions of hyphae a large mycelium could have, I realized something more surprising—that cumulatively, this could cover more distance than even the arctic tern who makes the world's longest animal migration, flying forty-four thousand miles or seventy thousand kilometers per year. It is often hard for us to imagine these mycelial journeys as we tend to un-derstand movement from a terrestrial point of view, as involving a rela-tively two-dimensional surface to traverse by foot or wheel. Mycelia inhabit a vast three-dimensional world in the soil, and at only one cell wide, they are so thin and fine that there is a lot of territory to explore; indeed, the scale is so different from our own that it is hard to imagine. When I first saw the claim that a teaspoon of soil might contain as much as a kilometer of hyphae, it seemed impossible, as a plant root in that space would only be a few centimeters. To put this into perspective, if you dug up a patch of soil a bit more than two meters by two meters to a depth of ten centimeters (half the length of a shovel blade) and placed the contents on a tarp and separated out all the hyphae, you'd have a long job ahead of you. Joined end to end, these hyphae could connect the north and south poles through the center of Earth, roughly eight thousand miles or thirteen thousand kilometers.[40] As the hyphae grow, they encounter a biodiverse and lively world, with many bacteria, insects, plant roots, fungi, and others to interact with. Some die from disease, some are eaten, and many connect with even more complex fungal networks through their growth and exploration.

In terms of eating, one can understand what fungi do in more active ways than merely "absorbing." As we saw above, fungi are mainly described as relatively passive decomposers, eating the dead; but as we saw in chapter 1, mycorrhizal fungi are critical to fostering the living. Around 90 to 95 percent of all plants are being nourished from youth to old age, and some, such as orchids, need fungi for seed germination. Fungi's vascular systems fuse with plant roots, and they receive carbohydrate sugars from their plant partners. In turn, they seek out food in the soil environment, sharing some with the plants. Two of the main forms of fungal eating are seeking out a mineral-rich diet, which could be described as "foraging" or "hunting."

Growing knowledge of fungal hunting abilities reflect how what we are able to see is inflected by the world-making qualities of those we watch as well as by the tools we use to mediate these encounters. For years, microscopes used a hot-burning light bulb that slowly cooked their subjects mounted on slides: this was a daunting environment for some forms of microscopic life. As scientists have changed the design of microscopes, they can see this underground drama unfold. They watch, captivated by the effective hunt of a fungus, a kind of organism rarely regarded as a hunter.

"Hunting fungi," such as the oyster mushroom (*Pleurotus ostreatus*), seem to have the most animal-like behaviors, which come closest to what are regarded as normative examples of active and agentive movement. *Pleurotus* hunt for live prey, especially for nematodes, which are microscopic worms that, like fungi, are almost everywhere. In this way, the tables can be turned, for nematodes often eat fungi: some nematodes have stylets that pierce hyphal walls and then suck out the insides of the fungus.

This reversal of the positions of predator and prey challenges more conventional understandings of what you may have learned in high-school biology, where organisms are placed in a rigid pyramid of trophic levels: with plant producers on the bottom, plant eaters in the middle, and meat eaters on top. We were often taught these levels are fixed. Looking at the relationships between fungi and nematodes, however, opens up another set of possibilities. It's as if sometimes a deer eats a wolf, and sometimes a shrub engulfs a rabbit. One of fungi's most obviously (to humans) agentive actions is when they not only fashion their own snares, but rapidly cinch them tight when a nematode moves through the trap. In a more patient and yet still cunning fashion, other fungi fashion a net or sticky trap. Others manufacture spiky balls that lie in wait, like caltrops, to penetrate the skin of a hapless nematode, amoeba, or rotifer. The fungi then, aware of their wounded prey, come in to finish the deed. Other fungi take a Trojan horse approach, creating spores that are ingested and germinate within the stomach of an unlucky nematode.

In chapter 1, I described how fungi actively drill into minerals, which they seek out selectively, not randomly. For example, they are attracted to grains of phosphorus-rich sand. Fungal hyphae don't just constantly exude digestive chemicals; they detect mineral prey and use powerful enzymes to dissolve these miniature rocks. Mycorrhizal fungi are effective in finding these nutrients and may provide their partners with up to 80 percent of the nitrogen and 100 percent of the phosphorus plants need to grow.[41]

Fungi don't always possess a single generic digestive enzyme; rather, some recognize different kinds of food and respond accordingly with specifically tailored batches of powerful enzymes to deal with different foods.[42] These enzymes dissolve materials before the fungus imbibes its meal. In other words, digestion starts taking place outside of their bodies. Whereas animals tend to ingest their food (bring it within their bodies for digestion), we could say that fungi "outgest" their food. After fungi convert potential food to digestible Food, they carry out what is called "active transport," moving materials from outside the cell membrane to inside. These actions are forms of work[43] and do not happen by mere passive absorption like a manufactured sponge. When you see a piece of fruit with mold in the middle and surrounded by a soft spot, these are digestive enzymes at work, you are watching someone else eat. Underground, fungi break down minerals that nourish and build their own bodies, as well as make them "bioavailable" to plants. Recently, botanists were surprised that when plants were given human-made fertilizers, created specifically for plants' needs, they could not uptake these fertilizers directly, but required fungal mediators to bring the fertilizer into their bodies.[44]

Sometimes fungal enzymes are but part of a series. When certain fungi eat a particular insect, for example, they create a number of different enzymes to get past the various layers of the insect's shell.[45] These are created in a distinct sequence; that is, the fungi can detect exactly which enzymes are necessary for each layer, and wait until each stage is complete before making the next enzyme. In this way, although a fungus that attacks insects is almost always described as a parasite, it could be viewed as a predator.

Conclusion

Anthropological studies have shown that scientists are of course not immune to the cultural ideas that surround them. This should seem obvious, but scientists have often portrayed themselves as the one social group with the special status of being, because of their scientific objectivity, separate from their own society. As historian Jessica Riskin shows, by the 1930s, anti-Romantic biologists promoted a form of writing that evacuated all forms of agency from the nonhuman living world.[46] Riskin's account explores how this dynamic came about, defining agency more broadly than the more human-centered versions as "an intrinsic capacity to act in the world, to do things that [are] neither predetermined nor random."[47] I, like a growing number of scholars, share Riskin's sensibility to broaden agency beyond humans, but yet not make it too sweeping.

As shown in this chapter, however, it is not always that all agency is, in practice, banned from scientific accounts of the natural world. There are members of what I call a "minor tradition" who advocate for retaining some sense of liveliness in the organisms they study. As well, some of their peers show how, in part, the strong tendencies (even among biologists) to evaluate diverse forms of life against an animal standard means that they tend to regard nonanimals as having diminished agency and capacities. An increasing number of scientists are researching the amazing capacities of plants and fungi in these regards, but these scientists face the threat of being labeled as romantic and unscientific.[48] Nonetheless, there is a massive appetite for such studies within both the scientific community and the lay public, and many are keen to discover that these organisms have many more abilities than previously documented.

Among best-selling books on these topics, Peter Wohlleben's *The Hidden Life of Trees* (2015)—a very different account from Peter Thompkin and Christopher Bird's (1973) *The Secret Life of Plants*, which, while quite popular, was slammed by many scientists—is full of anthropomorphizing language that goes further than I (and many scientists) feel comfortable with, and yet it is grounded in peer-reviewed studies that relatively few scientists would challenge. The valorization of trees is a long-standing genre in Western thinking, yet the valorization of fungi is fairly new. Just as I was finishing this book, Merlin Sheldrake's ode to fungi, *Entangled Life*, came into print and caused much excitement,[49] joining hugely popular TED talks on the powers of mushrooms. Within the academy, social scientists and humanists recently created "plant studies" (a listserv called "critical plant studies" began in 2019), but "fungus studies" has not yet emerged as a distinct group. Some scholars, however, have begun to engage with fungi to open up new realms of thinking.[50]

As discussed above, the notion of niche construction is one way to imagine animals as having a role in shaping their environments; this subverts the usual expectation that environments shape evolution and that animals adapt to the environment. This more common framework does two things: first, it separates animals from their environment, and second, it turns the environment into an active force that shapes animal evolution. In terms of the first implication, I suggest that organisms are always part of their own environments. As we saw in chapter 1, biotic life played a major role in shaping the planet. Every organism's actions, in some small way, play some, often seemingly insignificant, role in shaping its surroundings and yet cumulatively may create massive changes; for example, scientists have discovered the surprisingly powerful role that the breath and gas of millions of cows plays in changing the atmosphere.[51] In terms of the second implication, this framework understands one party as active and the other as passive, which, as I will show, makes it harder to see coevolution and to notice the diversity and importance of intimate connections, including symbiosis among mycorrhizal fungi, bacteria and animals, and lichens. As well, niche construction studies have been almost exclusively focused on animals, and I wish to extend those inquiries further, in terms of not just how animals, plants, and fungi are shaping their environments, but also how they experience their worlds.

To do so requires asking a larger set of questions about what it might mean to take a lively approach to understanding any kind of organism. In the first two chapters, we've seen how fungi, as a kingdom, have been so influential in making this world possible, a planet that is now chockfull of plants, fungi, and animals. We've mainly looked at how fungi have acted: how they formed relationships with plant roots, learned to break down wood, and so forth. Now, I'd like to expand the picture a bit, and bring us into questions of more so-called interior worlds, of how particular bodies exist in the world through continual acts of perception and interpretation.

CHAPTER THREE

Umwelt

The Sensorial Experience and Interpretation
of the Lively World

"How do you know but that every bird that cleaves the aerial way is not
an immense world of delight closed to your senses five?"

So marveled William Blake two centuries before we had the tools
to confirm that, at the very least, every dog is a world of delight closed
to our limited powers of sensorial perception. Out of such seemingly
simple discoveries across the animal kingdom sprang the rattling
realization that our notion of "reality" is really a plurality of radically
divergent impressions, shaped by the singular biases of perception that
each of us brings to our experience of the world. The same sliver of
"reality"—a table, a flower, a city block—is experienced in a wholly
different way by a bird, a dog, Blake, and you.

**—Maria Popova, "Diane Ackerman on the Secret Life of the Senses
and the Measure of Our Aliveness," 2015**

After returning from my first stint of Yunnan-based matsutake fieldwork
in 2013, I began my quest to find lively accounts of fungi. One helpful
clue to this quest came from someone who was not a mycologist but
was the son of one, and who himself became an anthropologist; that
person is Tim Ingold. His work, in turn, introduced me to the innova-
tive scholarship of Estonian biologist Jakob von Uexküll (1864–1944).
Uexküll came up with an exciting program of research that explored the
perceptual worlds of a wide variety of animals, which I later realized was
part of a "minor tradition" in biology, one piece of what some call

"whole organism biology." The time I spent with Yi mushroom hunters had inspired me to consider insects not as brainless drones but as creative agents in the world, who possessed senses often far superior to my own, and ones that might surpass my own five senses. They helped me appreciate insects' ability to be observant and clever hunters who are perceiving and interpreting the world. In chapter 4, I will describe how these mushroom hunters experimented with testing different ways to hide young and not-yet-picked matsutake from their insect competitors; the Yi hunters, aware that these insects had a point of view, also saw insects as observant learners.

I was intrigued by the possibility of better understanding how insects experienced the world *from the insects' point of view*, which, as it turned out, is an approach that was already inherent in Uexküll's project. Before my exposure to Uexküll, I read a paper by the American philosopher Thomas Nagel entitled "What Is It Like to Be a Bat?"[1] Nagel states some similarities between us and bats: we are both mammals who care for our young, and a bat's mode of echolocation has some parallels with our ways of seeing the world. Yet it turns out that Nagel's main point was that commensurability was impossible; he used this example to argue that humans can never grasp the mental state of fellow humans, let alone bats or other animals. Uexküll, on the other hand, was genuinely interested in knowing how bats, and different species of bats, perceived the world. I agreed with Nagel that humans can never completely know the bat world, but like Uexküll, I was eager to learn what I could about how other animals experienced their worlds, not only for that knowledge in and of itself, but also because I was interested in taking a further leap into imagining a fungal perspective.

Many have been impacted by Uexküll's work. Although he was influenced by German Romanticism, even for many scientists who were critical of such Romanticism, his work opened up a world of possibility. He became important for critical figures in my own field, such as Tim Ingold, as well as a number of scholars such as the German philosopher Martin Heidegger, the French philosophers Gilles Deleuze and Félix Guattari, and the Italian philosopher Giorgio Agamben. Uexküll's main contribution was the concept of the *umwelt*, which literally means "environment" in German, but which he used to describe the world as it is sensorially experienced by particular organisms.[2] Umwelt is the subjective environment that each species perceives and creates, shaped by their distinctive sensory apparatus. We often refer to "the five senses" as a well-established fact and then ask how other animals experience sight, sound, taste, smell, and touch; but more recently, this sensory box is opening up. Neurologists now recognize at least nine human senses;[3] and biologists are also finding more senses, some radically different from what we have imagined.

Uexküll asked: How do different organisms interpret the world around them, and how do they communicate their interpretations to others? He believed organisms possess complex forms of communication, and he wanted to understand them.[4] Uexküll's methods and interests influenced many of the researchers who later gained a profound understanding of the dances of bees, the songs of birds and whales, and how insects use pheromones to communicate.[5] Many of these realms of communication had been hidden to people until recently, in part because many didn't expect such sophisticated means of communication. Uexküll's notion that insects, for example, interpret the world and act as semiotic organisms motivated scientists to devote substantial attention to understanding how they communicate.[6] Uexküll was also interested in how animals use perceptual cues to interpret the world, and how they learn. Thus, he sought to understand the world of many animals long dismissed as lacking a sense of their surroundings. He conducted a wide range of experiments that offered clues as to how they understand and act in the world.[7] Importantly, his thinking recognized a great diversity of ways of being in the world, which neither put humans at the top nor endorsed the doctrine of human exceptionalism.

A Dynamic Umwelt Is Formed through the Accumulation of Active Engagements

My own world-making perspective emphasizes how an organism's umwelt is shaped through its encounters in the particular place in which it dwells. Here, I follow Ingold's understanding of the umwelt as not a fixed sensorial apparatus but as an active engagement.[8] This conception of the umwelt understands sensing as the means by which organisms, including humans, don't just *passively know* but also literally *make sense of* the world through their diverse senses.[9] Like some others, I see the umwelt not as "a touchable and tangible category, but rather an array of subjective and perceptive elements."[10]

Building from this notion of the umwelt, we can consider four main points that help us understand world-making in a deeper way. First, organisms experience the world in profoundly different ways, according to each organism's body morphology, senses, and attunements. Second, we can pay a different kind of attention to the particulars of how humans see, smell, hear, and touch the world once we accept that the human umwelt is just one of many umwelts (*umwelten* in German). Third, the recognition that some human senses are inherently limited in comparison to the senses of other kinds of organisms leads to the insight that there is an interpretive gap between human perception of the world and what one might call reality (or the world as it can be experienced by other beings). Put somewhat differently, one pitfall in not noticing the role of human perception is that we may conflate our way of perceiving reality with reality itself. Fourth, one of the more radical implications of Uexküll's approach is that this interpretive gap is not a human-only problem, but one faced by all organisms.

How the Human Umwelt Shapes Science

Let us briefly explore how human senses that determine our perception result in perception being conflated with reality and, moreover, affect how scientists unconsciously embed specifically human senses into scientific language. Various measures of the world, accepted as scientific standards, are shaped by the specific anatomy of the human body and, in particular, our species-specific senses. I suggest that if we don't unthinkingly accept this embedded perceptual bias—but instead pay attention to it and become more aware of our own positionality and limitations—we will recognize that humans are able to directly perceive *only* a narrow and particular sliver of life. In turn, we can then create new terms, and subsequently new understandings, that are not based on anthropocentric standards.[11] For example, "visible light" refers to the slice of the light spectrum that humans can see; "microscopic" refers to objects too small for the human eye to register; and "ultrasonic" refers to sounds higher than the range of human hearing.

In his book *What a Plant Knows*, botanist Daniel Chamovitz argues that scientific inquiry is shaped not only by anthropocentrism but also by a less discussed form of centrism: animal centrism.[12] He shows how biologists have tended to define perception in overly animal-centric terms. For example, he points out that the typical scientific understanding of sight conforms to the definition in *Merriam-Webster's*, that is, "the physical sense by which light stimuli received by the eye are interpreted by the brain and constructed into a representation."[13] This is not only animal centric; it's also vertebrate centric, because it considers only organisms with a central nervous system, a brain, and eyes. Instead, he asks, how might we understand "sight" for organisms that have neither eyes nor brains, such as plants (or fungi)?[14] As Chamovitz suggests, plants "are removed from our traditional understandings of the olfactory world because they do not have a nervous system, and olfaction for a plant is obviously a nose-less process. But let's say we tweak this definition to 'the ability to perceive odor or scent through stimuli.'"[15] Following Chamovitz, we can see how many organisms without a central nervous system, including plants and fungi, can detect chemicals in the air or water. Like seeing and smelling, let us now briefly examine the sense of sound, where new challenges emerge.

Having raised the problem of anthropocentrism, and animal centrism (especially a tendency toward vertebrate centrism), we can likewise consider anthropo-exceptionalism, whereby human senses are taken as the only ones that count. Let's consider sound, for as historian of science Sylvia Roosth argues, anthropo-exceptionalism has constrained what counts as "sound" itself. Replying to sound studies scholar Jonathan Sterne, who states that "sound is a product of the human senses and not a thing in the world apart from humans,"[16] Roosth says that such a stance

turns a deaf ear to those vibrations that are inaudible to humans yet are nonetheless key sensory capacities for nonhuman animals: ultrasonic vibrations among dolphins, bats, and dogs, for example, or infrasonic vibrations with which whales and elephants can communicate and anticipate danger in their immediate environments, such as earthquakes and tsunamis. . . . To productively draw sensory studies into conversation with multispecies science studies requires that the nonhuman umwelt be examined as rigorously and on the same footing as the human sensorium (and, indeed, to query the very notion of a singular and homogenous "human sensorium" in the first place). A more capacious understanding of sound could consequently reorient its focus away from not only anthropocentric but also "earcentric" models of sonic perception in favor of an extracochlear modality that recognizes entire percussing bodies as vibratory sensory apparatuses.[17]

I would also challenge Sterne's notion that "sound is a product of the human senses" by remembering that human-made and human-heard sounds acoustically overlap with sounds made and heard by many other species.

To use a vibratory metaphor, Roosth's approach resonates deeply with my own: an effort to come up with understandings of perception that are not based exclusively on the human body (actively dismissing all other bodies) and that explore how our human umwelt has shaped how we experience reality itself. I would like to expand beyond Roosth's interest in animals to consider how plants and fungi might sense vibrations (which is how we experience sound). There is some preliminary work, for example, on the new field of plant bioacoustics, which is new because for years scientists dismissed the very possibility. Scientists now argue that some plants are affected by the sounds of insects eating their neighbors[18] and increase nectar production after hearing the sound of pollinating insects,[19] and that underground roots might detect and move toward the sound of moving water.[20] In turn, we know that some plants are sound makers.[21] The field of fungal bioacoustics has barely begun, but we have one report showing how yeast changes its metabolism in response to sound.[22] I expect that with the right device, we could hear fungal sounds, as their hyphae push through soil, drill into rocks, strangle their animal prey, or react to being eaten themselves.

How Perception Shapes the
Experience of Time

One of Uexküll's surprising insights is that perception not only affects our experience of the auditory, visual, and olfactory world around us, but it shapes our experience of time itself. Enlightenment notions of time tend to portray it as a steady backdrop of life, like a universal ticking clock that is unidirectional, affecting all things equally. One of the best-selling books that draws on Uexküll's insights on perception and time is Alexandra Horowitz's *Inside of a Dog: What Dogs See, Smell, and Know.*[23] Horowitz is deeply immersed in dog research and uses the notion of the umwelt to open up new questions. She shows how dogs experience the world through noses, eyes, and brains, processing surroundings quite differently from the way humans do, but prioritizing smelling above all else. As Horowitz explains, smelling the world is quite different from seeing the world, in part because olfaction has a different relationship to time. Scents can be ephemeral or surprisingly enduring. They are affected by rain, wind, and materials like trees, grass, concrete, or metal that absorb these smells at different rates. For dogs, a breeze can expand their perceptual universe, as wind can bring a sense of what lies ahead. As Horowitz puts it, for dogs, scent gives them insights into "not just the scene [as] currently happening, but also a snatch of the just-happened and the up-ahead. The present has a shadow of the past and a ring of the future about it."[24] Thus, a dog's eyes and nose can create textures of temporality that humans are usually incapable of experiencing (because visuality is more constrained by the present).[25] The way dogs smell a world into being allows for a different experience of time, a kind of temporal layering of knowing what happened in the smellable past and what is carried on the breeze of the up-ahead. This is fundamentally different from how I was taught to conceive of time as three separate realms: past, present, and future. Thus, a study of other species' sensorial worlds may stimulate a broadened understanding and awareness of the role of specific kinds of perception in how we experience ourselves within the world,[26] including our experience of time.

It might not be surprising that our best understanding of other organisms' umwelts are based on studies of dogs, a human companion for more than ten thousand years. Compared to this familiar large mammal that many people have shared their homes with for years, it is much harder to imagine the questions to ask or the experiments to carry out to better understand how plants or fungi might perceive their worlds, including their experiences of time. This will be a challenging quest.

If we turn to insects, we can discover some hints. Compared to humans, the common house fly has an extremely fast processing ability. This processing ability might seem straightforward, but what is more interesting is how it affects a species' visual world. Visually, humans can notice up to fifty flashes of light per second. Chickens, however, notice up to one hundred flashes of light per second, a discovery that was revealed by noticing that chickens became agitated when new fluorescent lights were installed. Eventually, it became clear that while humans saw the light as steady, chickens saw it as a strobe light, pulsing at eighty strobes per second. In another example, films are calibrated on the human umwelt, which perceives film as virtually synonymous with seeing life itself, as continuous motion. When a fly looks at a film, however, it sees it more as a slow slide projection, with one still image followed by a black pause, and then another still image.[27] Thus, unlike humans, a fly would never confuse motion as represented in a film with the continuous motion of everyday life.

By shifting shutter speeds, humans can use film to temporarily modify their umwelt to speed up or slow down their perception of time. Film allows us to register and interpret actions that otherwise occur too fast to notice in detail or so slow that they exceed our patience (such as lying beside a mushroom for two days to watch it grow, or observing mushroom-insect encounters). As Oliver Gaycken suggests, "Time-lapse and slow-motion imaging are both realizations of the general principle that the cinema can dilate time; both procedures involve accessing temporal registers (otherwise) unavailable to the human senses via technical means."[28] In a crude way, slowing down a movie of life might better approximate the visual experience of animals like flies with faster-than-human motor responses. Yet finding out how plants or fungi experience time will require much more than speeding up a movie of life; it would require creative experiments to even partially grasp some of the dynamic ways that plants and fungi experience the world. Still, we can witness a number of time-related decisions that plants make. For example, Wohlleben describes in great detail the striking differences that exist between species of trees, and even those of the same species growing side by side, when it comes to the timing of certain decisions such as when to open leaf and flower buds, drop leaves, let a branch die off, and so forth.[29] These are actions that we can easily see, and yet they are rarely regarded as "actions," let alone a tree's "decisions." It is even more difficult to imagine how trees might experience time, but it is likely that "plant time" (which no doubt varies species by species to some degree) is related to light, at least in terms of the rhythms of the days and seasons.

I knew that plants could sense light, and so I imagined they had some kind of photoreceptor, but I found it surprising that compared to humans who have two photoreceptors (rods and cones), plants possess at least eleven kinds. We have just started to learn how different plants perceive light.[30] As usual, we know much less about fungi.[31] Scientists were intrigued to learn that some fungi have as many photoreceptors as plants. For plants, light is food, and for fungi it is something different, but we don't yet understand what it is for them, only that, in some ways, light helps mushrooms orient in time and space. Like our own skin, fungi create vitamin D when exposed to sunlight, so there are intimate relations to light. Fungi might use light as a clue for when mycelia have reached the air, a signal to knot together into a mushroom and to time themselves to grow at night when humidity is higher.[32] Indeed, fungi's propensity to grow at night and the amazing speed of their growth have influenced humans' reactions toward them—that they are uncanny organisms that appear out of nowhere.

Getting back to a human-specific sense of time, different senses are experienced in different temporalities. For example, although we think of touch as a direct experience, there is a gap in time between when something touches us and when we feel it. This gap can be noticed experimentally: tapping faster than sixteen times a second on a person's arm is experienced as continuous touch. There is a name for this gap, a "moment," a term we often use in a loose, impressionistic way but it is based on our umwelt, which is created according to the human nervous system. Anything slower will be experienced as tapping. Although we might think of plants as moving slowly, a human's sense of touch is relatively slow within the animal kingdom; many animals, such as the fly, not only see much more quickly but also feel much more quickly as well. Understanding how we compare with other organisms and learning about other senses that we don't have, or have but don't talk about, helps root our grasp of ourselves as one particular animal in a lively multispecies world.

Paying Attention to the Human Umwelt

As Jamie Lorimer notes, acknowledging the human umwelt quickly reveals the limits of our capacities: "Unlike most terrestrial mammals that communicate with pheromones, we depend on vision and privilege visual knowledge. . . . Human sensory organs make use of small portions of the electromagnetic, acoustic, and olfactory spectra for perception and communication."[33] Thus, all species, including humans, use their senses to actively interpret the world, and these interpretations play a role in making that world. Awareness of our own limited abilities, along with noting the capabilities of whales and bats, helped us develop technologies such as radar and sonar. When we better recognize and understand the role of our own senses, we might change our orientation toward the world. Instead of conflating our perception with the world itself, we become aware of how these senses shape the very contours of the worlds we make.

How does my training as an anthropologist affect my understanding of the umwelt? As an anthropologist, I learned that there are many ways to be human; and I was trained to see the world as full of human difference, including the fact that humans speak more than six thousand different languages. I was taught that each language carries a unique way of understanding and orienting to the world, and that these are profoundly different. In my graduate training, I was never told anything, really, about human perception as compared to other animals.[34] We considered animals only as "good to think with" in human terms, understanding their symbolic value.

From this perspective, I imagine that although many humans as a species share a similar umwelt, they are not exactly the same; rather, humans perceive the world in a range of ways. The most prominent way that anthropologists have explored perception is with respect to the role of language, which affects the senses. Anthropologists such as Asifa Majid might show us some of these connections, as both language and everyday performances shape the relative attention we pay to various senses. When Majid worked with the Jahai people of Malaysia and the Maniq of Thailand, she was surprised to find a rich vocabulary of dedicated smell words.[35] At first, I assumed that the Jahai might have hundreds of these words, like how the French have helped foster a seemingly endless collection of terms to describe wine, but then I learned that, at most, the Jahai had fifteen such terms. Even though the number seems low (although high compared to English, which, according to Ed Yong, are few: "There are only three dedicated smell words—stinky, fragrant, and musty—and the first two are more about the smeller's subjective experience than about the smelly thing itself"), the Jahai pay great attention to smells and frequently discuss their differing qualities.[36]

These cultural differences are not just fixed, however; within a given cultural milieu, perceptual orientations can change over time. Over the past few centuries, Europeans have increasingly focused attention on the visual to the deficit of other senses. Such a move was not accidental as, according to Constance Classen and others, Westerners created a hierarchy of the senses: associating visual and auditory modes with reason and rationality,[37] while dismissing smell, what philosopher Immanuel Kant once described as the "most ungrateful . . . and the most dispensable" of the senses.[38] In turn, since Europeans created the vast majority of scientific instruments for exploring the world, it is no surprise that they mainly focused on enhancing sight (with such technologies as microscopes, telescopes, and X-rays). Even technologies that use vibration (like radar, sonar, and ultrasound) or smell (like gas chromatography) convert machine-made senses into visual representations, thus reinforcing and increasing our visual reliance.

This kind of slow drift of the senses may occur, but there is also evidence that people's senses, such as smell and hearing, are trainable within a generation, which informs my sense that the umwelt is dynamic and not fixed. Alexandra Horowitz, mentioned above, has been able to train and greatly enhance a person's sense of smell and use it to know her surroundings more actively, to experience the world a bit more like dogs do.[39] Other human examples include how the blind learn to perceive the world through echolocation. Daniel Kish talks about how, like many blind children, he learned to make clicking sounds to hear his surroundings, but was discouraged from doing so. He disregarded these admonitions and developed an acute ability of echolocation. At first Kish could walk without a cane and eventually become proficient at riding a bike in the city. His paper, subtitled "How to See Like a Bat," nods to Nagel's paper on bats and describes skills he teaches others and hiking trips he leads for the blind.[40]

In the examples above, we have heard about differences of perception within *Homo sapiens*, which may vary between groups, over centuries, or within a few years. I cannot imagine that humans are unique in this regard—or that they are exceptional. Thus, I want to be clear that I don't assume a kind of species universalism whereby all members of a given species are the same, even in such seemingly "hard-wired" (to use a mechanistic metaphor) aspects as our own bodies. Such a presumption has been built into most claims of scientific knowledge; an experiment on one rat or rabbit is regarded as applicable to all members of the species. As we recognize differences among individual humans and among human cultures, we should be amenable to noticing differences among groups within the same species, to considering that other animals have forms of dialect, of culture. Such questions are already being actively explored.[41] A world-making approach sees bodies as generated through a set of encounters, and through active learning and change within a lifetime, as well as through the more usual scientific notion that bodies change over "evolutionary time."

This active way of seeing bodies made through encounters is quite different from how I saw Uexküll's work being used by many thinkers; I was often disappointed that the more lively implications of his work were greatly constrained. Why was this so? Although Uexküll wrote a lot, there is one main story that is recalled when referring to his work: that of the tick (*Ixodes rhitinis*). Here is Agamben's version:

> This eyeless animal finds the way to her watchpoint [at the top of a tall blade of grass] with the help of only its skin's general sensitivity to light. The approach of her prey becomes apparent to this blind and deaf bandit only through her sense of smell. The odor of butyric acid, which emanates from the sebaceous follicles of all mammals, works on the tick as a signal that causes her to abandon her post (on top of the blade of grass/bush) and fall blindly downward toward her prey. If she is fortunate enough to fall on something warm (which she perceives by means of an organ sensible to a precise temperature) then she has attained her prey, the warm-blooded animal, and thereafter needs only the help of her sense of touch to find the least hairy spot possible and embed herself up to her head in the cutaneous tissue of her prey. She can now slowly suck up a stream of warm blood.[42]

This is a striking story, and Agamben's version is as good as any. In some ways, he created a more active rendition of the tick's life than I had heard before. For instance, the tick lands on an animal described as "prey," a term more associated with predators. Agamben used the female pronoun "she" rather than the less accurate but more commonly used and more alienating term "it." We can also see the semiotic emphasis on "signals" and "perception," which is a hallmark of Uexküll-informed biosemiotics. In Uexküll's rendering, the tick's umwelt has only three signifiers (things that matter to the tick's universe): the smell of butyric acid, the warm temperature of mammals, and the hairiness of mammals. It presents the tick's world as starkly impoverished, sensorially limited, and triggered by only a few kinds of sensorial cues. It is sometimes mentioned that a tick has lived without food in a laboratory for up to seventeen years in a coma-like existence. In Uexküll's account, the tick's life is described using engineering language, as if it were a robot activated by triggers that cause a fixed and predictable response.

Although Uexküll wrote this story to explain how an organism's umwelt constrained its world, it has, unfortunately, often become *the* story about the umwelt, reinforcing a view of nonhuman organisms as sensorially impoverished and limited to actions demarcated by a fixed repertoire of reflexes, which has already been commonly established in professional and lay biology. I realized that this example, often the only one that others knew from Uexküll's work, was almost opposite to an active world-making approach that I was trying to explore. Many of his accounts of other organisms are more alive with possibility, not only showing senses that may be far beyond human capacities, but also demonstrating an animal's active engagement with the world and a sense of improvisation. By drawing on and extending Uexküll's other research, I contend that organisms actively shape their umwelt through a continual engagement with the world as they experience and live it.[43]

Forms of Communication, Forms of Living

Uexküll's work was all but forgotten for a number of years. In 1976, the Hungarian-born, British-trained Thomas Sebeok was a semiotician and linguist at the University of Indiana when he found, in the library, an original German version of Uexküll's book *A Foray into the Worlds of Animals and Humans*. The funny thing was that he had already read Uexküll's work thirty years earlier and had dismissed it. That was when he was a graduate student at Cambridge, where he'd read a poorly translated English version. In English, he had found it "bafflingly murky," but when he read it in German it was "electrifying."[44] Sebeok steadily propagated Uexküll's work and created an academic field called "zoosemiotics." It was later called "biosemiotics" after some researchers started studying plants, fungi, and bacteria. Before Sebeok's interventions, the field of "semiotics" referred solely to humans' creation and interpretation of signs, such as spoken and written language. Building on Uexküll, Sebeok turned the anthropocentrism and human exceptionalism of semiotics into a multispecies version of semiotics that would enhance our notions of world-making.

Although the number of biosemioticians grew quickly after Sebeok helped reinvigorate the field, there has still been a strong animal bias, which follows general trends in biology. When the botanist Francis Hallé surveyed a number of biology textbooks, he found that at least 90 percent of examples were of animals; fewer than 10 percent concerned plants.[45] In my own studies of biology textbooks, fewer than 1 percent of examples looked at fungi. Given this animal centrism, I was thrilled to discover the books *Biocommunication of Plants* and *Biocommunication of Fungi* and surprised to find that both were edited by an Austrian philosopher, Guenther Witzany.[46] He was an unusual philosopher; his work had almost no references to Continental philosophy but was chock-full of peer-reviewed scientific articles. I read a lot about plants to help me better understand possibilities for fungi.

Witzany argues that plants have an extensive chemical "vocabulary" that functions "as signals, messenger substances, information carriers, and memory medium in either solid, liquid, or gaseous form."[47] Plants actively interpret signs, both biotic and abiotic, and Witzany suggests that they follow three kinds of rules, similar to human communication: participants interpret meanings through "syntactic (combinatorial), pragmatic (context dependent), and semantic (content-specific)" contexts.[48] He argues that air is often thick with communication, much of which is extraneous to the plants' livelihood—what we might call "noise." Plants can use more than twenty different types of chemical communication, and plant roots can produce more than one hundred thousand different chemicals.[49] Each of these chemicals can become a

form of communication, and each chemical reacts with others to create
novel signals. For instance, plants use chemicals to communicate when
they are wounded by insects. These chemical messages are received by
nearby plants, even those of different species, which in turn can secrete
chemicals that make them less appetizing to insects. Such signals are
not simple "all or nothing" alerts; rather, they differ in proportion to the
attack and can be subtle forms of communication.

Although humans began to discover some of these airborne signals
only as recently as 1990, it is likely that some grazing animals learned
that plants communicate through the air currents much earlier, changing their own eating strategies to reflect this. Other animals remain
oblivious to such phenomena, even to their peril. In one study, researchers observed that a number of kudu antelope fed almost exclusively on
acacia tree leaves. Wounded trees sent out alerts and nearby trees secreted poison. Some antelope ended up dying from the accumulated
poisons secreted by the trees. Giraffes, however, approached the communicating trees from downwind, somewhat like how we might think
of a lion stalking its prey. Giraffes ate their fill from trees that hadn't
smelled the alert and filled their leaves with defensive toxins.[50]

Plant communication differs in relation to the type of damage. At
first, I assumed that the chemicals produced by plants were automatic
responses to being damaged, almost like the smell of a wound. I later
learned that this was not so, for it matters if a leaf is eaten by an insect
or ripped off by a gust of wind. Plants attacked by insects tend to mount
a chemical defense and alert others. Plants torn by strong winds do not
produce protective chemicals. Such an effort would be without purpose
for there is no way to protect themselves in the moment against wind
damage, and somehow the plants recognize the difference.[51] Some
plants don't have a generic response to herbivores. Instead, they can
detect chemicals in different herbivores' saliva and thus can respond to
particular species.[52]

Witzany's work, inspired by Uexküll's approach, stresses that there
are always two aspects of semiotic communication: the creation of signals and the interpretation of them. I realized that I had previously
regarded the interpretation of signals, through hearing and seeing, as

innate. And while I had regarded activities such as running and hunting as acts more or less skillfully practiced, I had not imagined the act of interpretation as skilled. But in reading others influenced by biosemiotics, I came to see interpretation as "enskilled"—that is, it is not innate but a matter of embodied practice whereby skill can be gained—and thus some do it better than others. This concept was first used to describe humans[53] and more recently nonhumans.[54] It helps us understand the potential for all beings to increase their skills over a lifetime, like how birds take years to learn to build high-quality nests.[55] Thus, I see skills such as interpretation, finding food, dealing with predators, and so forth not just as innate and fixed ("instinctive behaviors") but as abilities that are dynamic, changing through practice, experience, and social learning. I also acknowledge that some individuals are better at certain things than others; some plants, for instance, are likely better than others at communicating through early predilections and experience.

For plants, ignoring signals or incorrectly interpreting them can have negative consequences. If plants produced protective chemicals in response to every false alert, they would soon use up their own internal resources and perish.[56] The insight that plants actively interpret the world through the smell of airborne chemicals and the taste of herbivore saliva challenges conventional assumptions that plants are basically passive and inert. Scientists have also recently discovered that plants communicate not only through air but also through soil. Underground, they are especially reliant on their fungal partners. In these underground forms of communication, networks of fungal hyphae relay messages, sharing information between species. The older understanding of fungal mycorrhizae was that it was a connection between one mycelium and one plant, which moved water and nutrients back and forth. Newer understandings reveal a vast underground communication network, connecting many different species of trees and fungi, and that it not just circulates water and food but also plays an important role in plant communication. These signals greatly assist forest trees in alerting each other to impending insect damage and drought; this information is vital to maintaining a healthy forest in a dynamic world.

These multispecies and cross-kingdom soil worlds were a surprising discovery that really gained ground only in the late 1990s. As mentioned in the last chapter, scientists had long known about mycorrhizae, but many assumed that the fungus was parasitic on the "host tree." Eventually, after it was established that there was *mutual* benefit, scientists still believed that the relationship was between one tree and one fungus that moved water and nutrients back and forth.

Three important discoveries led researchers to appreciate the profound ecological role of fungi. First, researchers found that multiple species of fungi did not attach only to a single tree but that they stitched multiple trees together into a network. Second, they realized that these underground networks not only distributed food and water; they also played a vital role in plant communication. This idea had been circulating for a while. In 1987, Grime and his colleagues found that in laboratory conditions certain plants could transfer carbon from a donor to a receiver using mycorrhizae. A 1988 paper by the botanist Edward I. Newman, "Mycorrhizal Links between Plants: Their Functioning and Ecological Significance," imagined the existence of a "mycelial network" linking plants.[57] "If this phenomenon is widespread," Newman wrote, "it could have profound implications for the functioning of ecosystems."[58]

In 1997, the journal *Nature*—considered one of the most prestigious academic outlets, usually reserved for cutting-edge, synthetic research with major implications—came out with a headline on its cover that captured the world's attention.[59] The main author of the article in question was Suzanne Simard, a fellow Vancouverite and professor of forest science at the University of British Columbia. Her coauthored paper had a more typical scientific title: "Net Transfer of Carbon between Ectomycorrhizal Tree Species in the Field." But the journal highlighted the research by drawing a compelling and catchy analogy between these plant and fungal networks and the Internet—and the analogy quickly spread: "the wood-wide web" (fig. 3.1) In more popular outlets, the "wood-wide web" soon replaced what had been an established scientific term: the "common mycorrhizal network" (CMN).

FIGURE 3.1. Simard's discoveries renamed "the wood-wide web" on the cover of *Nature*, August 7, 1997. Image reproduced with permission from Nature Publications.

Simard's work showed that nutrients could flow between different trees, including different species, including birch (*Betula*) and Douglas fir (*Psuedotsuga*). These flows were the most controversial aspect of Simard's findings in that these various species seem to violate the neo-Darwinist principle that all behaviors improve fitness for oneself and one's offspring—what some call the "selfish gene hypothesis,"[60] thus challenging the operating assumption of mainstream biology—that competition is *the* driving force that guides evolution. Quite the contrary, her work suggests that forms of cooperation may be incredibly widespread. As we saw after the discovery of mycorrhizae in chapter 2, for decades scientists assumed this was a parasitic relationship, in part because they assumed that interspecies relationships were overwhelmingly antagonistic. Studies by Simard and others looking at the wood-wide web used radioactive carbon as a proxy for the question of benefit (or at least a traceable and quantifiable unit to measure sharing). But as we will learn, this is only one part of the larger relationship.

In her studies of carbon sharing between stands of birch and Douglas fir trees, Simard found that nutrients seemed to flow back and forth between these stands over the years. One day, a colleague at Simon Fraser University, Andy Mackinnon, told me about a conversation he had with Simard about this phenomenon. He asked her, "Why would the trees want to do this? Why would they share with others, especially those that were not related?" Simard looked at him and chuckled: "Andy, Andy, Andy, what makes you think the trees are in charge?" This sentiment parallels insights from other researchers that show that almost all studies on the establishment of mycorrhizal relations have focused on plants' role in "orchestrating" the fusion,[61] thereby viewing fungi as passive.[62] In other words, much research on the CMN had been plant centered. Simard and other researchers, on the other hand, had moved from a plant-centric to a mycocentric experimental design to test for other possibilities, a focus that I also advocate.[63] Simard's sense was that fungi would redistribute nutrients based on whether it was a better growing year for one tree species or the other; by helping both species thrive, fungi had a better chance of hedging their bets regardless of the weather.

Almost all these studies, new and old, looked at how the CMN relayed a single element: carbon. Not only is carbon the new coin of the realm for critical questions about climate change and using trees to help store atmospheric carbon, but it was also one of the easiest elements to follow through the system because scientists could deliver radioactive carbon to plant leaves and follow it as it moved throughout plant bodies. The next step was moving beyond this carbon-centered approach to see how hyphae exchanged messages, sharing information between species, using what are called "infochemicals"—that is, chemicals without any food value. Some researchers argue that these signals greatly assist communication among trees, alerting them to impending insect damage and drought.[64] Such information is vital for forests to maintain their health in a dynamic world, for trees make many choices and carry out many actions based on this information.

How Fungi Dynamically Engage the World

In conversations with Chinese mycologists, I learned that a number of mycological experiments are inspired by botanical studies; botanists, in turn, often test out questions first raised with animals. Scientists show how fungi use active forms of embodied communication throughout their lives—such as establishing relations with their symbiotic plant partners, organizing their bodies to notice and ensnare nematodes, detecting and harvesting minerals, and so on. From a world-making perspective, we can see that fungi interpret their environment and respond to it with a wide range of actions. This challenges common scientific classifications of fungi as immobile, assuming that they just seem to "grow," not "move," and that fungi lack the capacity for behavior or action. As I mentioned earlier, within a broader European folk understanding, animal-like forms of mobility have been regarded as a key criterion for agency. According to this criterion, fungi, like plants, seem to be rooted in a particular place and (hence) lacking in agency, but as I showed in chapter 2, not only do fungi have the capacity for movement; in fact, they are constantly moving, but mainly underground where they are not noticed by people.[65]

Drawing on the study of biosemiotics, we can also understand acts of communication as forms of action and as forms of agency. Sometimes these acts are enduring and sometimes fleeting in relation to rapid change in the world around them. At every moment, it is likely that fungi are actively perceiving and responding to the dynamic world around them. Making a small but important pivot, we can say that their acts of growth, their performance of living, are part of this changing landscape: they are not just reacting to the world; they are part of it. This way of understanding a world-making perspective emphasizes organisms in their environment—in contrast to seeing organisms as isolated and independent beings.[66] Biological approaches like niche construction theory also emphasize the organism in the environment.

One of the best-known examples of fungal communication with animals relates to the truffle. Truffles emit certain chemicals only when their spores are ripe, and they release a powerful scent as a kind of communication; and depending on the species and soil, some of these chemicals can move through a meter of soil. This form of communication might seem like a one-way affair, but it becomes a two-way dynamic when animals dig down to find the truffles, eat them, and then later disperse truffle spores in a new place.[67] This scent is not just the smell of a delicious food; it is chemically reminiscent of flying squirrels' sex pheromones.[68] Some of these squirrels, in turn, depend almost entirely on truffles for food and eat little else. This could potentially be the world's most valuable diet, as well as one of the most intensely flavored: as my friend Willoughby asks: Can you imagine a diet consisting almost solely of food that smelled like sex?

Other truffles reproduce by attracting insects. In the case of the tiny truffle beetle (*Leiodes cinnamomea*), only the female is attracted to the truffle's scent. Females burrow underground, and the males, in turn, are attracted to the female's scent. Underground, beetles of both sexes consume their subterranean feast, coming back to the surface and spreading truffle spores as they defecate.[69] In other words, *Leiodes* don't only have a species-specific umwelt, but to some degree have a sex-specific one, in terms of what chemicals attract them. This is a matter of survival for truffles, because without *Leiodes* and other animals to transport their spores, truffles would have quickly become extinct and would never have evolved. These interactions are just one small part of the world of fungal engagement.

Semiosis is important, but it is not the only thing going on. I found that a number of researchers influenced by Uexküll tended to focus almost exclusively on perception as a mental process, rather than explore more broadly how different bodies inhabit the world and create worlds through their messy and material liveliness. To explore the lives of fungi beyond a semiotic-only frame, I dove into the mycological literature to learn about how fungi interpret and interact in various diverse relationships: with tree roots, bacteria, a wide range of other fungi, underground insects like nematodes, and aboveground ones like sciarid flies. To view such diverse bodies in relationship can help move us beyond a focus on how organisms think—a semiotic-only approach—toward an understanding of how organisms live, how their bodies are engaged in many interactions, such as having sex and dying, eating and killing.

Let's look briefly at several bodily interactions, where fungi seek out insects and turn them into food as well as into vehicles to spread their spores. One of the most famous cases is the "caterpillar fungus" or "grass worm" (cordyceps), which thrives in high-altitude grasslands in Southwest China. This fungus parasitizes the caterpillar of the ghost moth, which is turned into a "zombie." The caterpillar starts crawling upward in the soil, but stops an inch below the surface. It then dies in the perfect position for the fungus's last move. Next, the fungus spreads throughout the caterpillar's body, consumes its internal organs while leaving its exoskeleton intact. Finally, it creates a mushroom that sprouts from the caterpillar's head, which pops out aboveground and spreads its spores with the wind.

In an even more complex set of relations, one of cordyceps' relatives is *Beauveria bassiana*,[70] which uses two kinds of insects to spread itself. First, it parasitizes a caterpillar and then produces volatile chemicals that attract mosquitoes; the mosquitoes, in drinking the caterpillar's fungi-containing blood, become infected and are themselves subsequently turned into fungal agents.[71] Fungi can also take over plant morphology and behavior, creating "zombie plants" that transform plant bodies through fungal action. Likely the most common example is *huit-lacoche* (the Indigenous Nahuatl term from Latin America), known in English as "corn smut," for this fungus grows on corn seeds, inducing them to grow into huge and almost comical shapes, which become filled with spores.[72] Other fungi reengineer flowers so that instead of pollen they produce fungal spores.[73] As one parasitologist explained to me, such fungi take the plant's "baby energy" and use these nutrients to serve their own reproduction.

Beyond their use of smell, fungi attract insects in a number of ways through shapes, color, and phosphorescence. For many years, fungal phosphorescence was assumed to be a mere metabolic by-product, but in the 1980s mycologists started to wonder if it was used by the fungi to signal insects. They found that fungi do not glow much in the day but increase their luminescence at night, possibly to better attract spore-carrying insects.[74]

Why did it take so long to investigate this? One reason is that many Western mycologists assumed that fungi are passive. Although botanists view flowers as the means by which plants attract pollinators, less attention has been paid to the possibility that the colors or glow of fungal fruiting bodies could be attracting pollinators (or what should more properly be called "sporinators").[75] This discovery was likely delayed because of the human umwelt, in that humans tend to be diurnal as opposed to nocturnal, so we know far more about daytime ecology than nighttime ecology, based on our habits of activity and sleep and the simple fact that the human eye does not see well in the dark.

Thus, when humans search for mushrooms, they are late to the party compared to other species over the evolutionary long haul, and also humans are often late in the actual season. Even before they are fruiting, fungi are already engaging with many interlocutors. Mushrooms, like plants, are also interpreting the world, listening for signals from others (more like an active form of chemical detection, like smelling but without a nose). They are acting based on communications from others. Yet as we saw earlier with animals and plants, Witzany argues that fungi—whether they are communicating at the intracellular level (to manage their own growth and defense), or communicating with other members of the same or other species—must actively *interpret* these chemical-born messages.[76] They are likely constantly receiving many signals: the air is full of them, and fungi are attuned to some and cannot even perceive, let alone interpret, others. They use context-dependent clues to help decipher meaning. They also create their own signals, some of which are in a chemical form that they diffuse into the atmosphere and through the soil matrix. Others they broadcast through an underground mycelial network, the "wood-wide web."[77] As mentioned earlier, while this wood-wide web was previously described as a kind of telegraph system for relaying messages between plants, we are now realizing that such fungal networks are much more than inert infrastructure. Indeed, fungi and plants built this web together, and it is growing and decaying as they make new relations and old connections break down.

It is obvious, then, that mushrooms are not merely passive objects, just waiting to be eaten. Instead, mushrooms, like animals and plants, are semiotic beings, creating signals and interpreting those they pick up from others. Fungi are also more than semiotic; they do more than think and communicate; like all organisms, they are embodied in their world-making. Underground or within other substrates like wood, mycelia are constantly seeking food and mates, dealing with disease and predation, deciding where to explore and when, if at all, to knit themselves together to make the mushroom that humans—and many others—might notice and come to find.

Conclusion

Uexküll's theories help challenge the premise of human exceptionalism; I deeply appreciate how they motivate a curiosity about the lives of all species. Although he focused on animals, had he been alive today, I would not be surprised if he were keenly interested in work on the sensorial abilities of plants and fungi. How might we selectively embrace and enrich the insights of Jakob von Uexküll to gain better understandings of multispecies world-making?

His work has several radical implications: (1) all species have their own unique worlds that are informed by their bodies and their perception capacities, (2) all species must interpret their perceptions, and (3) there is a gap between perception and the world.

By helping us recognize that we humans—as bipedal mammals with a limited sensorial repertoire—also have an umwelt, Uexküll reminds us of our limitations and our tendencies to interpret the world through our specific senses. The recognition that our nervous system (compared to a housefly's, for example) affects how we experience action, and therefore time, is surprising to many. It shows that we don't experience the world *as it is*, but only as it is experienced in our specific body. We are compelled, in turn, to recognize that there is a great diversity of bodies, and thus a great diversity of ways of reckoning the world. Uexküll's work helps us to question the notion that time exists in a standard and universal way that is detached from life itself. This contrasts with what Isaac Newton claimed about time and is generally accepted in science. Uexküll offers us a sense of situated time, which exists only through the experiences of particular embodied organisms.

We have just recently begun to gain a better understanding of how the human umwelt shapes our attunements toward, interpretations of, and encounters with other organisms. In terms of sound, it has taken us a long time to notice the communications of animals and plants that use sound waves above or below the human threshold because we first had to create devices to hear those sounds. In terms of sight, it has taken focused effort to attune ourselves to bees that dance in their dark hives or to forms of microscopic life.[78]

This is not only about the limitations of a human umwelt, however, for distinct conceptual frameworks guide scientific thinking among the majority of scientists, who tend to restrict their questions and thus the experiments they carry out. Let me just rehearse a few of the elements of this dominant conceptual framework that have made it more challenging to see fungi as world-makers. First, forms of animal centrism inflect what is recognized by scientists as examples of movement, behavior, learning, and agency, which have historically been defined in terms of vertebrate animals.

Second, many scientists have worked hard to eliminate anthropomorphism (imparting human sensibilities onto nonhuman beings).[79]

In many ways, I support this effort. But the notion of anthropomorphism is sometimes applied too widely, with the assumption that such a broad range of abilities and actions are available *only* to humans. Thus, this overly restrictive definition, based on assumptions, can end up fostering the less appreciated problem of human exceptionalism, which contends, a priori, that only humans can possess certain qualities.[80] I am hopeful that an active reading of Uexküll's insights can help us move beyond a mechanistic framework to show not only that other organisms make occasional decisions but that their lives, like ours, consist of continual and never-ending acts of perception and interpretation.

Uexküll's work fostered the sensibility that other beings have their own ways of knowing the world, and from this, their own forms of communication. By adopting a model that moves beyond "the five senses," and that attends to registers beyond human capacities, we can begin to learn what other species are saying.[81]

Although plants and fungi operate in very different ways from how animals do, they still must perform a daily quest to meet their needs (water, soil nutrients, sunlight, sex) in dynamic environments to which they must adjust and in which they respond to others who are intimately interested in them in some way (including trying to eat them); plants and fungi use their senses and the range of actions in their own repertoire. While much work remains to be done, we are starting to learn more about the surprising ways that fungi engage the world through their own umwelt.

My active reading of the umwelt stresses how organisms come into being in relationship to each other's presence. This reading may differ from Uexküll's famous and striking metaphor of each species existing in its own autonomous soap bubble—a metaphor that gives a strong sense of the umwelt as trapping each organism in a particular and limited perceptual universe, a self-contained world, but that fails to show how species continually engage with each other. For organisms to survive as a species, they must connect their lives with the lives of others—as predators, prey, and parasites; as flowers and pollinators; as sources of spores and agents of the spores' dispersal; as coinhabitants. In other words, while Uexküll saw that umwelts (*umwelten*) actually overlap and intersect, I believe he paid insufficient attention to how organisms are in a state of becoming through their relationships with others. Building on Uexküll's work, I move from the wide range of fungi addressed in the previous two chapters to deeply explore the realm of one species—the matsutake. In the chapters that follow, I explore how matsutake have moved throughout the world, always in relationship to others: to certain trees and then as part of a global commodity chain. Exploring matsutake's umwelt and its interspecies relations is one step toward better understanding multispecies world-making, with and beyond humans.

CHAPTER FOUR

Matsutake's Journeys

The potential for matsutake to grow in Southwest China began with a collision of tectonic plates that occurred about fifty million years ago. This crash created the Himalayas, the tallest mountain range on Earth. Some imagine the movement of these plates as a relatively slow process, where one plate just slips under the other. But sometimes these plates don't slip but collide like massive sumo wrestlers. Just sixty million years ago, long after the dinosaurs were gone, India was an island that lay fully south of the equator. Within just ten million years, it had sailed northward, crossing a vast expanse of ocean, and rammed into the Asian plate. Now, the northern tip of the Indian plate almost reaches as far north as Philadelphia, Madrid, or Beijing, at forty degrees north latitude; and it's still traveling north.

Before the crash, India was relatively flat and warm, as was the Asian landmass it collided with. The crash elevated rocks high enough to almost enter the stratosphere (more than twenty-two thousand feet), forcing clouds to release their moisture. The Himalayas are one of the world's "water towers," where precipitation built up for millennia and eventually created a vast land of glaciers, which some call the Third Pole (after Antarctica and the North Pole). Over the millennia, these glaciers shaped the valleys that created lands fit for trees, matsutake, humans, yaks, and many others.

—Reflections on fieldwork

The first half of the book looked at how the kingdom of fungi shaped our planet, how the everyday actions of various species of fungi constitute what I am calling world-making, and how employing a lively version of the umwelt might help us gain insight into the perceptual world of different fungi, at least by guiding questions and motivating curiosity.[1] In this second half of the book, we explore the life of one particular species complex—the matsutake mushroom—as it becomes entangled with other fungi, plants, and animals in a particular part of the world. We travel to China's Yunnan Province in the eastern Himalayas, which is located in the country's southwest corner, bordering Myanmar/Burma, Vietnam, Laos, the Tibetan Autonomous Region, and Sichuan Province (see plate 2).

These seemingly insignificant mushrooms power a vast economy spanning the globe, but as we have already learned, they are much more than a preexisting resource for human profit and pleasure. Matsutake do not just exist in the world for humans as an object of the hunt, a commodity in the basket, and a meal on a plate; rather, they are living beings carrying out their own life projects, with specific forms of liveliness. This chapter looks at how matsutake are building worlds in Yunnan, with and without humans. First, it shows how matsutake spread across Asia, building relationships with trees, insects, and others. Next, I show the rise of the matsutake economy, and how matsutake became increasingly connected with Yi and Tibetan pickers in Yunnan—becoming *the* mushroom to find and sell—and then with Japanese buyers and scientists.

In a vast trade network connecting nearly a dozen countries, Yunnan is now the world's richest source of matsutake mushrooms. Unlike the circulation of commodities such as oil, lumber, and steel—which make their way to nearly every country in a complex network—transnational flows of matsutake are more like a wheel and set of spokes: the mushrooms travel almost entirely to just one country in the hub: Japan. To make this trade happen, a million people around the world coordinate their actions every day during the season—finding this mushroom in the forest and then sending it to urban centers where, in turn, it will be shipped by air to Japan, then auctioned off and distributed to thousands of outlets, from large grocery stores to small greengrocers.

The story of the matsutake trade can easily be told as a human-centered tale in which matsutake appear as a readymade commodity, entering into a trade totally dominated and controlled by people. But there are other ways to tell this story. A number of pundits talk about matsutake "production" (as if it were produced in a factory), or they refer to it as an "agricultural good" (as if it were grown in a field). Yet it is easy to recognize that historically matsutake were neither produced nor grown through human intentionality. Whereas a plant like corn is deeply tied to a history of intimate relations with humans who decided which seeds to eat and which to save to plant next year, matsutake has its own history that far precedes human actions. Corn arrived in Yunnan only two or three centuries ago—carried in first by Spanish traders from Latin America alongside potatoes, chili peppers, and tomatoes; matsutake, on the other hand, has likely been growing in Yunnan for millions of years. In Latin America, corn's ancestors have lived for thousands of generations, producing corn together with hundreds of generations of humans. As corn travels around the world, it continues to diversify as it adapts to local climates, soils, insects, and human desires. Corn and human lives are now so entangled that corn relies completely on humans for its own reproduction. Unlike wild grains that shatter when ripe, corn seeds remain trapped within a tight husk and their seeds do not dislodge: few corn seeds on the soil can grow into a mature plant without human assistance. Matsutake, on the other hand, can thrive without human intervention, although humans have sometimes unintentionally expanded matsutake's potential habitat because people fostered vast areas of pine in one way or another.

Thus, while the actions of people picking matsutake are necessary for the economy to come into being, the fundamental basis of the economy is matsutake's own presence, which happened without human effort; and this presence is often taken for granted. In other words, this trade relies completely on the ability of matsutake to grow, move, and survive—that is, on its liveliness. Matsutake moved across the world thanks to their own exploratory travels. Sometimes they traveled long distances via spores blowing in the wind; shorter spans were accomplished on the feathers of birds and fur of animals; and sometimes they hitched a short ride within the guts of small insects. As matsutake explored, they didn't do so on their own, but always in relation to the diverse world around them.

Matsutake on the Move in Deep Time

Exactly how and when did matsutake come to the place that became known as Southwest China's Yunnan Province?[2] The specifics of this journey will almost certainly remain a mystery. Yunnan is but one spot on the globe where this species complex can be found. Figure 4.1 is a speculative map of matsutake's nomadic travels over deep time (as we saw in figure I.3), now marking Yunnan's approximate location.

These journeys began many years ago, long before most crop plants arrived in China from the Middle East or the Americas and long before modern humans entered this landscape. Through their efforts and relationships with trees, they made a living in this place.

Viewed from space, matsutake territories look like giant irregular footsteps or islands, which are separated by vast distances. How did these islands come into being? Matsutake likely spreads both by air and by land. Overall, most spores land close by the mushroom. It is a rare event, but sometimes a spore will travel far and create an island of territory. Self-directed movement by mycelia, on the other hand, generally creates contiguous territory. Although such movement is slow, over deep time, much is possible. We don't know how fast matsutake mycelia can spread over such time periods, but some old and large mycelia (like *Armillaria*) grow about one meter a year.[3] At that rate, it would take matsutake 2.5 million years to grow the twenty-five hundred kilometers from Kunming to Northeast China, which is the other place in China that matsutake are found.[4] If matsutake spread mainly by mycelia, then perhaps some habitats within these vast areas changed, causing matsutake to disappear between most places but leaving behind some islands (fig. 4.2).

In this region of the world, the island of matsutake includes parts of Yunnan, Sichuan, the Tibet Autonomous Region of China (TAR), Nepal, Bhutan, and likely a bit of India. From this island, the next substantial island of matsutake to the east is in Northeast China, where it

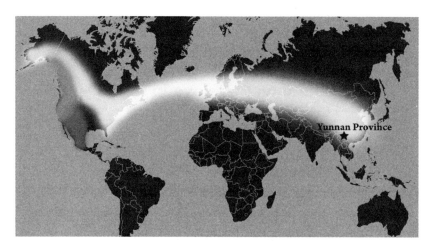

FIGURE 4.1. A map showing the possible spread of matsutake species with Yunnan highlighted on the map. With kind permission from Andreas Voitk.

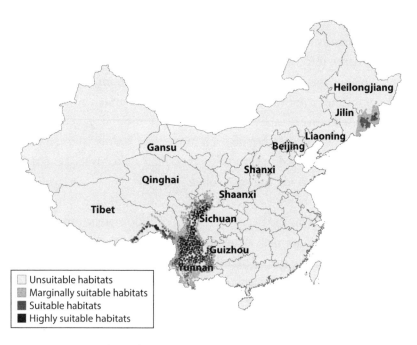

FIGURE 4.2. Survey data and GIS algorithms describing four kinds of potential habitats for matsutake in China. From Y. Guo et al. 2017.

extends into North and South Korea.[5] From there, matsutake jumps a body of water to proliferate in Taiwan and Japan, and it then leaps across a much larger body of water (the Pacific Ocean) to create a large zone of proliferation along North America's west coast, from Alaska to central California, and also creates another island in Mexico. They established themselves in a few islands along the US and Canadian east coast, and then jumped the Atlantic, appearing in isolated islands in places such as Finland, Turkey, and Morocco.

Returning to Yunnan and shifting perspectives from high up in space to the view from a black-necked crane (one of my favorite Northwest Yunnan birds), we see that what looked like a contiguous island of matsutake habitat is actually a series of smaller, separate islands, as this terrain creates many microclimates and microhabitats. The high peaks above tree level are far too cold and windy for matsutake; at this southern latitude they prefer an elevation between twelve hundred to thirty-five hundred meters.[6] Sometimes they live in a single elevational ring around an isolated peak. Another major factor that influences tree and matsutake growth is the orientation of the slope, especially whether it faces north or south. The slope's angle to the sun plays a major role in shaping local microclimates, especially in the higher elevations. South-facing slopes are sunnier, warmer, and drier, whereas north-facing slopes are shaded by the mountain itself, thus becoming colder and wetter. Matsutake fruit in response to a series of cold snaps in the fall, influenced more by the temperature of the soil than the temperature of the air. In turn, trees and their shade influence the timing of a particular patch. Sunnier places cool later in the fall. Trees, in turn, shape the soil's microclimate through their discarded leaves and needles, and the amount and kind of this forest duff insulates the soil from the air: in terms of temperature, more duff delays the matsutake's fruiting season.

For matsutake, certain trees are not merely sources of shade and duff but are, more importantly, critical life partners. In other words, matsutake never grow alone, out in a grassy field; they grow in forests where they form intimate and obligatory relations with one or more of these trees.[7] Yunnan's matsutake are able to fuse with either pine or oak, whereas in Japan, matsutake are mainly associated with the Japanese red pine (*Pinus densiflora*)—a relationship that was the basis for its Japanese name: pine mushroom. In fact, among the many Tibetan names for matsutake, the main one is *beshing shamo* (often shortened to *be sha*), which means "oak mushroom," for matsutake seem to prefer oak in the overlapping territory of the mushroom and Tibetans. Thus, after matsutake arrived in Yunnan, whether by mycelia or by spore, they settled down and began searching for their potential tree partners. They could not live long in places where these trees were missing, or if their mycelia or spores could not find these trees, or if the trees refused to form mycorrhizae with them.

Making Relations with Trees

Some fungi are able to form relations with many different plant partners. Matsutake, however, are often choosier, sometimes having only one partner in a given location. Around the world, the species complex is known to partner with more than a dozen tree species in more than a dozen countries, and these numbers of partners will probably grow. One reason is that mushroom hunters will likely discover more partners; but there is a second, less appreciated reason: as you read this chapter, matsutake are expanding their range of possibilities, experimenting by making relations with new kinds of plants even as they may lose old ones.

It is well appreciated that many partnerships between particular fungi and particular plants evolved over deep time, and it is impossible to pinpoint the origins of such pairings. As I mentioned in chapter 1, we have evidence of mycorrhizal relationships fossilized in a piece of chert that was found in Scotland and dates to about four hundred million years ago; so this strategy of pairing with plants is an ancient one. For matsutake, we don't know when they evolved and coevolved with a series of different plants, but this mutual becoming has likely been happening for hundreds of millions of years.[8] Yet, we can also see how new plant-fungi relations can happen quickly, maybe even within a decade or less. This phenomenon becomes visible when we notice how "invasive mushrooms" create relations with plants they have never encountered before.[9] Shifting our sense of timescale once again, it is little appreciated that each potential relationship between a specific tree and specific mycelium must happen in real time as well.

Although pine and oak trees have partnered with matsutake to share food, many other organisms seek out matsutake as a source of food or a place to nurture their young. For some organisms, matsutake is just one kind of food among many; but for at least one plant species—the candy cane—matsutake seems to be the only thing it eats. Thus, when humans hunt for matsutake, they are joined by others who vastly outnumber them and who are far more experienced at finding their prey, including millions of insects, birds, and mammals. These hunters are likely driven mainly by smell and sight as they search for indicators of matsutake's presence.

As described earlier, fungal perception and communication are largely a chemical affair, but scientific knowledge of how these work in dynamic ways is just beginning.[10] Matsutake produces a kind of baseline odor, a spicy cinnamon-like smell. The most frequently quoted description of this odor is by David Arora, who calls it a "provocative compromise between 'red hots' [the sweet, spicy candy] and dirty gym socks."[11] For some humans,[12] this smell is one of its key appeals and for market hunters a key quality that distinguishes it from other white, similar bodied "look-alikes," some of which are poisonous, that may grow alongside matsutake. The chemical qualities of matsutake provide signals that travel through the air and soil and are detected by plants, animals, and fungi among others, luring in some species and likely repelling others.[13]

Yet it is not just that fungi use chemical communication to announce their presence, as a fixed and consistent signal. As I mentioned in chapter 2, fungi use chemicals in a dynamic way, in relation to specific encounters. Fungi respond to chemicals that move through the soil and produce their own messages. Mycelia and tree roots use initial chemical signals to find each other within the dark soil. In certain situations, they sense and grow toward each other, but as they come into closer proximity, there are likely other sets of chemicals that each plant and fungus produces, which create more specific conversations and decisions regarding if and when to fuse. In some situations, they decide to fuse, whereas in other situations they pass and go their own way.

As a particular tree is engaging in an exchange with this new myce-lium to facilitate making connections, the tree might already have my-corrhizal relations with dozens of other mycelia. It is hard to say where the mycelia begin and end in these underground networks (the "wood-wide-web" discussed earlier), which challenges our usual understanding of trees or fungi as isolated individuals. Likewise, when mycelia are looking for a mate, they follow six possible stages that are similar to their quest for a compatible tree: searching, finding, exchanging, touching, fusing, and mating. They search the soil by means of chemical signals, find each other, and then exchange another set of substances. Next, they may touch, but as with their intimate relations with trees, touching does not always result in the act of fusion.[14] Finding other compatible mating types does not guarantee an act of mating; mating is not automatic. For many species, fungal mating eschews human intention and escapes sci-entific prediction. As well, not all forms of fusing result in mating, which is defined not as the fusion of bodies but as the sharing of DNA—that is, the fusion of cell nuclei. Some fungi fuse bodies but never exchange nuclei, what scientists call "infertile somatic compatibility," showing us, again, that fusing is not always the same as mating. In most animals and plants, these two different events can occur almost instantaneously in a series of events called "fertilization," but in fungi these events can also be separated, such that the mycelia fuse but potentially the nuclei do not fuse for centuries.[15]

For matsutake, traveling by spore and making a new home is not a singular stroke of luck, for mating is required to produce another gen-eration. It seems that in the history of the planet, there are some rare cases where a single pregnant animal (such as a bee or a turtle) has come to a new land and been able to create a whole population from her own belly. The same has happened from seeds produced by plants that are "self-fertile;" their flowers are able to pollinate themselves and produce fertile seeds that can spread. Yet as I described earlier, mushrooms differ from plants. A spore is not like a seed, and neither is it like a grain of pollen. Fungi have created a wondrous variety of lifestyles, and some create asexual spores that can produce a new version of themselves with spore-making powers—a kind of self-fertile microclone—but matsutake spores might require sex with another to make more spores.[16] A matsutake spore can germinate and grow into mycelium, but unless it finds another of a different mating type and exchanges its genes, it cannot produce mushrooms that can, in turn, make spores and spread through the air to new places.

Whether or not they mate, matsutake mycelia can make relations with different plants to exchange minerals and water for photosynthetic sugars; indeed, they must. They can carry on an active life by spreading underground. When it comes to mapping the presence of matsutake, we have usually relied purely on finding their fruiting bodies not their mycelia, as scientists have rarely tested soil for its DNA presence. Thus, matsutake are likely far more abundant than we have realized because mated or not, mycelia can grow for many years without ever fruiting.[17]

Although a number of animals are interested only in the fruiting bodies that matsutake makes, there are some animals and plants that are happy enough to eat unmated mycelia; and nematodes and other small underground animals are often nibbling them. Sometimes the mycelia survive, if they grow fast enough to outpace the rate of being eaten, or, as I will describe shortly, turn the tables and eat their potential predator. Because mycelial bodies don't have central organs, they are less vulnerable to being killed, compared to, say, animals with vulnerable body designs that include arteries or a heart.

Matsutake has one plant predator that we know of, with the Latin name *Allotropa*, known in English as the "candy cane." It has developed a particular and even exclusive craving for the taste of matsutake. *Allotropa* likely hunts using matsutake's distinctive chemical signature. Like matsutake, it has crossed both sides of the Pacific Ocean, making a living on its western (mainland Asia's east coast) and eastern (North America's west coast) shores. *Allotropa*'s presence always means the presence of matsutake mycelium; it is one of the only reliable visual signs, and one that can easily enter the human umwelt, which is more attuned to sight than to smell. When young, *Allotropa*'s stalk is red with white stripes in a beautiful candy-stripe pattern, and it contains small white flowers that have a red center. The fact that it has no green is a bit of a giveaway that it does not create food from the sun; *Allotropa* eats from others, matsutake's own water and minerals as well as the sugars of the mushroom's life partners, the trees. *Allotropa* stalks are often eaten by animals when the plant is growing, but after death its stalks can persist for three years, making it a long-lasting sign of matsutake's underground presence.

Sometimes human hunters can also learn to detect the particular aroma of matsutake mycelia. A friend of mine in Vancouver says that every season she uses *Allotropa* as a guide to "tune" her nose: digging back the soil around the plant, she discovers the white and sometimes even bluish patch of thin cotton hyphae, gaining a kind of "search image" for her nose. But she reminds me that matsutake are almost always growing in the presence of many other kinds of fungi, each with their own odor. Thus, in contrast to walking through a forest that contains only matsutake, where she can search for its unique smell, she must attune herself to a symphony of odors released by a fantastic variety of fungi, each with its own bouquet. The challenge, then, is to pick out matsutake's own distinct note among many fungal notes. A matsutake hunter soon learns that this particular mushroom attracts so many other organisms that a hunter is rarely the first to discover each mushroom; instead, they hope to find it before too many others have started to eat it or have eaten it for too long.

In the same way that *Allotropa* indicates matsutake's underground mycelia, mammals such as elk or deer can also be guides for those humans who recognize that they, too, can lead the way to finding matsutake. For example, in Oregon, elk or deer will sometimes paw the ground and unearth buried "mushrumps," kicking them out[18] and thereby alerting human hunters. In Yunnan, I did not hear of any large animals with a distinct craving for matsutake, and there is almost no scientific information on this subject, since very few researchers study the role of fungi in animal diets in Yunnan or elsewhere.[19] It is likely, however, that many of these organisms were already eating these mushrooms for many millions of years before modern humans arrived.

Insects, however, are matsutake's main animal predators; and they are looking not only for food for themselves but for places to lay their eggs. A mushroom can become a home and a source of food for their growing children. Thus, every day, humans are late to the scene in finding matsutake before various insects have already discovered them.

When matsutake get ready to fruit, their mycelia grow in a thick ring called a *shiro* (meaning "white castle" in Japanese). This ring, somewhat like the truffle's *brûlé* (meaning "burnt" in French), dramatically transforms the soil in the ring, making it appear "bleached, dry, [and] hydrophobic (water repelling) . . . a site of intensive chemical weathering of the mineral substrate."[20] Many think that matsutake is "weakly competitive" against other fungi and creates this zone of exclusion around itself, exuding acids and antibiotic secretions to rapidly gain nutrients and keep others at bay.

Environmental History: People Enter the Landscape

Like the story of matsutake's movement to this part of the world, the story of humans' arrival in Yunnan is also complex but still little known. Waves of human ancestors, starting with *Homo erectus*, probably began to arrive more than a million years ago.[21] Alongside other animals, humans might have eaten matsutake and spread their spores through their feces, but unlike the relations between truffles and certain animals, little is known about matsutake's relations with animals, such as how different ones act as matsutake spore vectors or spore destroyers.[22] What is likely the case in Yunnan, however, in terms of deep time, is that for a long while, humans barely mattered for matsutake lives.

From matsutake's perspective, the main way that humans came to matter was probably through fire. *Homo erectus* learned to make fires to keep warm, cook food, and then, eventually (and most importantly for the sake of matsutake), to transform landscapes. Fire was a helpful but dangerous ally, aiding humans in flushing and driving wild animals. Early humans might have also noticed that after a fire, grasses and shrubs grew with additional force, providing more graze for the large mammals they hunted and creating sites of attraction. As human-made fires joined lightning-made fires as a powerful force for changing the landscape, this likely facilitated the spread of pine forests, as some pine seeds sprout only after a fire and often grow quickly on burned-over land. In places like Yunnan, more fires meant more pine trees, which in turn meant more matsutake potential.

The last wave of people, modern *Homo sapiens*, may have arrived here only sixty thousand years ago or less, but the archaeological record is sparse until about four thousand years ago.[23] In some parts of China, the climate cooled over the past five thousand years, from when elephants roamed the area of present-day Beijing, which may have happened in Yunnan as well. If so, this would extend the range of pines and oaks down from the high edge of the Himalayas to the lower elevations and southward. Beginning around two millennia ago, northern-based empires—led by Mongols, Manchu, and Han—carried out military campaigns to bring Yunnan under imperial rule. As the human population increased, people cut down forests for building materials, to fuel smelters, and to clear land for crops.[24] As we will see in chapter 6, yak-herding groups in northern Yunnan also burned pasturelands, which could kill tree saplings. Compared to those in eastern China, in Yunnan, human populations have been relatively low. During some periods, such as when wars raged in the mid-1800s, many trees began growing on abandoned farms, which increased potential matsutake habitat. Elsewhere in China, logging was more intensive, and rulers established large-scale tree plantations for trees such as *Cunninghamia*, which have so far refused relations with matsutake, thereby shrinking potential habitat.[25] In Yunnan, repeated logging tended to concentrate near large towns or mines where populations were more stable and trees were in great demand for fuel and building supplies, and here the forests were not likely to spring back unless these places were abandoned.[26]

In the 1930s, a series of important events led Yunnan to rapidly build road and airport infrastructure that would later prove critical for its matsutake economy in the 1980s. Ironically enough, these events were precipitated by the Japanese themselves when Japan invaded eastern China. In response to Japan shutting off supplies to the east coast, the Burma Road was built to bring supplies from the west. In 1938, nearly two hundred thousand Burmese and Chinese laborers accomplished a massive engineering feat: they connected Kunming to Lashio in Burma, a distance of 717 miles (1,154 kilometers), using only hand tools such as shovels and hoes.[27] Although Yunnan had been a crossroads for more than a millennium, the Burma Road was the first major road not made for human porters and their pack animals, but for vehicles with wheels. Long-existing paths used stairs to allow the hoof and the foot to gain traction on steep terrain, but such designs were antithetical to travel by wheel, and all the steps were converted to slopes.

The fight also took to the air, when Japan launched bombing raids into Yunnan. An airport was quickly built in Kunming and was used by a unit of American pilots called the Flying Tigers. They flew from India over "the hump" of the Himalayas to bring supplies. Their planes were not designed for such altitudes, and, in frequent white-out conditions, many crashed.[28]

Yunnan experienced not only a boost in physical infrastructure, such as roads and airports, but also a massive influx of intellectual infrastructure. In the 1930s, when Japanese troops took over east coast cities, a large group of students and professors (some of whom would later win the Nobel Prize) fled China's elite universities near Beijing. Over several months, they walked to Kunming and created a new university, Lianda. After the war, some professors stayed and created Yunnan's first nationally prestigious academic centers, such as the Kunming Institute of Botany (KIB), which became a key national center for fungi research. Before World War II, residents of northern China saw Yunnan as a distant, disease-ridden outpost on the frontier of barbarians, a place for the emperor to exile difficult people. Yet by the end of the war, KIB quickly rose to prominence, and its researchers have played important roles in fostering the matsutake economy, studying matsutake's ecology, economy, and potential uses as a medicine.

Now, let's turn to what was happening in Japan at the time. You may recall that historically matsutake was a sumptuary good, and there were laws against it being eaten by commoners: this was considered a food fit only for the imperial court.[29] Poets had already described its mouth-watering smell more than twelve hundred years ago,[30] and it was regarded as "a taste of autumn from the old days."[31] The body of the matsutake—its qualities of smell, its appearance, its texture in the human mouth—all combined to make it a distinct mushroom that in Japan is freighted with deep layers of historical and cultural resonance.[32]

After centuries of being a luxury good, matsutake suddenly seemed to proliferate in Japan in the late 1930s, just as Japan's troops were landing in China. In 1940, matsutake were so abundant as to become cheap for the first time in Japanese history; Japan's wartime victory seemed likely, and plans were underway to attack Pearl Harbor within a year. Japan's allies, the Germans, were devastating county after country in Europe. It turned out, however, that both military success and matsutake proliferation would soon falter. The matsutake season of 1940 was not part of an upward trend but a peak, and matsutake levels began to plummet. Japan's emperor surrendered in 1945, and by the 1970s, domestic sales had dropped from over twelve thousand metric tons to under two hundred metric tons.

Matsutake's precipitous decline in Japan was seen as a national tragedy, and many tried to explain why the mushrooms were disappearing. Biologists discovered that forests were in decline, and it took scientists another decade to find that acid rain—coming mainly from Chinese coal-powered factories and energy plants—was damaging trees.[33] Thus, Chinese activities played an ironic role in the transnational matsutake trade. China's coal smoke was unwittingly and unintentionally boosting its own role in the matsutake economy by generating pollution that reduced the abundance of matsutake in Japan itself. This prompted the Japanese to undertake two actions. First, in a series of actions that might be surprising today, Japan did not just blame China for acid rain but poured massive funds into retrofitting Chinese factories.[34] These efforts resonated with Japan's role as China's most important trading partner and source of international development aid in this era.[35] Second, Japanese entrepreneurs teamed up with scientists to find new sources of matsutake around the world, which included traveling to Yunnan. It still remains somewhat of a mystery why they thought to look there, but rumors circulating in Japan suggest that the hunger for matsutake prompted the scientists to improve some of the world's early GIS systems, which used computers to overlay maps of climate, soils, forest types, and so forth to predict places around the world where matsutake might be growing. Whether informed by GIS or not, in many cases the scientists' guesses were correct, and entrepreneurs and others started building a global trade in matsutake.

This new trade significantly changed sales in Japan. In the early 1970s, almost every matsutake eaten in Japan was gathered domestically, but by 1980, Japan imported 362 metric tons of matsutake, making import levels roughly equivalent to domestic levels. Soon after, imports increased dramatically, and within nine years, imports had increased almost sevenfold to 2,210 metric tons.[36] It was not that already existing markets in other countries were redirected to Japan; rather, Japanese scientists and businessmen traveled abroad to search out new matsutake populations, and they began the economy from scratch. Almost nowhere in the world, except in Korea, was there an existing economy in place.[37] In North America, Canada began shipping matsutake in 1978. Some pickers reportedly earned $1,000 in a week, even with the relatively low value of about ten Canadian dollars per kilogram.[38] In the same year, two Japanese researchers mapped matsutake's distribution in the United States and Canada, revealing new places that pickers did not know about[39] and encouraging the development of a North American market.

There are different origin stories for the beginning of Yunnan's matsutake trade.[40] I heard several times in Lijiang that Yunnan's matsutake were first brought to international attention in 1987 by a team of visiting Japanese scientists, who, posing as butterfly collectors, were actually searching for matsutake. They discovered it growing on the side of Jade Dragon Snow Mountain (Yulongxue Shan) and filled their sacks. Coming off the mountain, they were stopped and questioned by police and forced to reveal their true purpose.

In Japan, however, I was shown an article published in 1981 that indicated an earlier expedition by Japanese scientists. Three researchers—Yasuto Tominaga, Ryoko Arai, and Toshio Ito—traveled in search of Yunnan matsutake. They were joyful after they found it growing in Chuxiong, Dali, Lijiang, and Diqing, and they wrote a Japanese account in the *Hiroshima Agricultural College Bulletin*.[41] Their article does not reveal how they established contacts with the Chinese, but it is nearly certain that they relied heavily on Chinese guides. In fact, Diqing remained off-limits to foreigners, so traveling there may have been quite difficult. The researchers found that trade with Japan already existed but at a low level. The mushrooms were driven to Kunming and flown to Beijing and then proceeded onward to Japan, likely by ship. Word about China's matsutake quickly spread in Japan, as there are strong links between dealers and scientists, and Tominaga was one of Japan's prominent researchers.

During the late 1980s and early 1990s, Japan experienced an economic boom, and matsutake prices shot up. At that time, in places like Canada, there were reports of pickers making a $1,000 a day. As a friend in Vancouver said, "In the late 1980s, almost no job paid that well; not even the CEOs were earning that kind of money. Plus, it wasn't even really work, just picking mushrooms in the woods." These kinds of profits were unheard of, given that other people working in the forest gathering salal stems for the florist trade, or even wild chanterelle mushrooms for the European market, never made money like that. To put this in comparison, in 1992, on average, the highest-grade matsutake fetched nearly ten times the price of chanterelles ($75 a pound vs. $8 a pound). But matsutake prices are also subject to extreme variability within a single season: in one year, the price for top-grade matsutake moved between $40 and $520 a pound![42] Even within a single night and sometimes over a few hours, the price would double or fall by half.

In China, such levels of wealth were never achieved; no one made the equivalent of a thousand Canadian dollars in a day. Yet in proportion to previous incomes in Yunnan—often less than $200 a year—this was a kind of bonanza for pickers. As one old Tibetan man put it, "It was like we found a new kind of gold, but we didn't have to dig for it. It came back every year, even if we picked every one the previous year. It was not like anything we had before."

Why didn't pickers in China earn the kinds of fortunes possible in North America? It is likely that Japanese dealers knew that they could pay Chinese pickers much less for these mushrooms, thereby keeping more profit for themselves.[43] As well, it was still relatively difficult to bring them to market in Yunnan. Mushrooms often traveled without refrigerated trucks, on rutted dirt roads or bumpy cobblestone roads for many miles. One dealer I talked with, who had suffered great losses after his boxed mushrooms were bruised by the journey, thereby losing much of their value on route to the city, ended up hiring villagers to stand in the back of the truck with a large basket of matsutake on their backs—the thinking being that their bodies would cushion the bumps, making up for the poor roads. It did help the mushrooms a bit—they arrived in better shape—but the people holding them up arrived in rough shape. The dealer acknowledged that the work was grueling, and few were willing to make repeated overnight trips to the city.

In the 1980s in central Yunnan, an older Yi matsutake hunter told me that, before the Japanese market emerged, matsutake were picked mainly by children who wanted to make some spare change. But later in the decade, word got around that the Japanese very much wanted this one particular mushroom that local people had been gathering but had not especially relished. At first, people from Kunming came north to buy fresh matsutake and then brined them in plastic barrels, which were sent to Japan by boat. Matsutake can be kept this way, but the price for these pickled mushrooms is low. One of my friends in Kunming, Tian Shitao, who had worked as a nurse for extremely low wages, said that she was the first in her work unit to go into the pickled matsutake market. She took an overnight bus to Lijiang, bought up all she could, hauled the heavy barrels to the bus, and then came back to Kunming and looked for dealers to ship them abroad. International shipping was very difficult at the time, encumbered with lots of bureaucracy and challenges. She might spend up to forty-eight hours on a bus, leaving Friday to travel over bumpy roads and returning on Monday, worried her boss might find out, but happy to make a bit more money.

In the late 1980s, matsutake prices rose, and plane fares dropped. Boxes of fresh matsutake were flown to Japan, opening up a new era in the global matsutake trade. With this, "matsutake fever" was spreading in Yunnan, though its peak was short-lived. It was not that people had picked all the mushrooms, but that the prices fell. By 1991, after a five-year period of high profits and conspicuous consumption of luxury goods, including matsutake and tailor-made suits, Japan's boom economy turned out to be a bubble economy: it burst.

Since then, matsutake prices have oscillated frequently but overall have gone down substantially, and pickers abandoned their fantasies about becoming rich until the year 2017, when prices shot up again, in part because few matsutake fruited that year. As I describe in chapter 6, some far-reaching government policies that began in 1999 made logging illegal, and logging revenues had funded 80 percent of local government budgets. These laws created a crisis, shifting the overall economy away from cutting trees and toward picking mushrooms, unintentionally boosting matsutake as a key activity in the regional economy. Matsutake hunting emerged as one of the easiest ways to make money and one requiring the least amount of investment in training or tools; it was an activity done largely according to one's own schedule, and with no boss. By the year 2000, China was selling more than half of the world's matsutake exports to Japan, and Southwest China made up around three-quarters of China's exports.[44] Despite the fact that matsutake was not always generating high prices, many villagers embraced hunting and selling it and other mushrooms during the season. From 2004 to 2006, the price was about twenty yuan (US$2.50) per kilo, but people still picked.[45] However, in 2017 the price shot up to three thousand yuan per kilo, spurring another case of "matsutake fever" that saw people hunting with great excitement. As I describe in more detail in chapter 6, few matsutake appeared in 2017, indicating that it was likely the regional scarcity that affected global prices.

At the other end of the food chain, Japanese consumers make several demands concerning the quality of matsutake, which significantly shape how the chain functions. They view matsutake as wild, not cultivated, and this means that they expect very low levels of contaminants like pesticide compared to what they allow for farm crops. This demand forces Chinese dealers to be extremely cautious, for there are many ways, besides intentionally spraying the forest or warehouse, that matsutake can be unintentionally exposed to pesticides. It can happen via drifts from nearby fields, or from carrying them in baskets that previously held sprayed vegetables, or by touching them with unwashed hands after working in the agricultural fields. Perhaps even more important, unlike European consumers, who expect to eat their porcini and morels in dried form, Japanese consumers for the high-end market demand that matsutake be fresh, not frozen or dried, both of which processes change the smell and the texture and sully the pristine, just-picked-from-the-forest quality of matsutake. One dealer who had traveled to Japan said the best analogy for matsutake is fish for sushi: the fresher the better.

This demand for fresh, versus dried or frozen, mushrooms places significant demands on pickers and exporters. Dried mushrooms can be stored for years, making them a commodity that can be stockpiled while waiting for a better price. But fresh matsutake can't sit long; a pile of valuable mushrooms will quickly turn to mush. Thus the fast pace of the trade, the metronome to which everyone moves, is set by the quickly decaying quality of matsutake-with-insects and the culinary demands of Japanese consumers. Such demands for freshness might change, for not all Japanese consumers are gourmands, and few want to spend exorbitant amounts of money on matsutake; this is evidenced in the fact that one can easily find instant noodle packets with artificial matsutake flavor. But given Japan's centuries-long passion for matsutake, there will likely continue to be willing buyers, at a fluctuating price, for some time to come. Among scientists in Japan, even those who are working hard to try to cultivate it, there exists a fear that if they are successful, the end product will be different, and it will not be as good. Somehow, it will not have the wide range of special, ineffable qualities that make wild matsutake a food endowed with somewhat mystical properties.

Conclusion

The market, we see, requires four sets of actors, not just the usual two that most studies of commodities focus on: producers and consumers. On the producer side, regional markets need enough pickers who are able and interested in devoting months of their yearly cycle to finding matsutake in the forest and hurrying them to market to make it worthwhile to ship them to Japan. If everyone moved to the city and no longer wanted to do this work, the market would end; but these days, Yunnan has one of the largest concentrations of active mushroom hunters in the world.

On the consumer side, the market is fueled by enough Japanese interest in buying these mushrooms. If everyone suddenly lost the desire to eat and gift matsutake, the market would end, but Japanese interest in and demand for matsutake remains the highest in the world. In Yunnan, probably millions of small matsutake are picked every year that cannot be shipped to Japan because there are now laws in place that forbid sending these ones abroad.[46]

The two other groups that are critical to the market are, of course, not humans but the trees that are matsutake's critical life partners and the matsutake themselves. There was no guarantee that matsutake would start living in this place, and the journeys that brought them here were likely compelling adventures. Matsutake have been living in Yunnan since deep time, and it is even possible that Yunnan is the place of their evolutionary origin; scientists continue to inquire into this, tracing relationships and lineages that may be buried in the code of their DNA.[47] Human presence played a role in matsutake's territorial presence throughout the region, yet not always a prominent one. Taking the perspective of deep time, fire has been more important than the ax or the plow in shaping matsutake landscapes, and as I explained above, human populations have ebbed and flowed. World War II really put Kunming on China's map in a critical way and boosted the infrastructure such as roads and an airport that would prove crucial to the rapid growth of Yunnan's matsutake economy. Even though Kunming grew quickly, in much of Yunnan, rural populations remain low. In Tibetan areas, for instance, much of the land has only about twelve people per square kilometer, and often there are more yaks than people.

In Yunnan, as in many places around the world, it is likely that the majority of matsutake appear and disappear without entering the human market at all; most are never picked by people but are nibbled by a range of other beings. Pickers notice the rasping teeth marks of a slug, the missing pieces caused by a bird's beak, or the boring of the fungus gnat (*Diptera*), which uses the mushroom as a nursery to raise her young larvae, which are easily distinguished by their black eyes.

There is no guarantee that matsutake will continue to flourish in Yunnan; perhaps, as was the case in Japan, they will start to vanish. When

I talked with villagers in the two main groups that hunt matsutake in Yunnan, the Yi and the Tibetans, I found that many of them did not take matsutake's continued presence for granted. Instead, they saw them as a gift—one that might disappear. Since the 1980s, when matsutake appeared, each group has started to recalibrate their annual patterns of work in relationship to this mushroom, in turn reconfiguring their relations to other living beings that have their own demands, such as corn and goats, barley and yaks. These matsutake hunters remain grateful for the gift of matsutake that has appeared in the forests surrounding their homes and which has now become one of their own critical life partners. The mushroom provided them with levels of wealth they had not seen for more than fifty years and with new possibilities, as well as some difficult choices.

CHAPTER FIVE

The Yi and the Matsutake

Late July and mushrooms are everywhere: in the woods, on the street, and in the markets. Many are "just mushrooms" to be dried and ground into powder, mainly to add flavor to food in Europe, a "new spice trade." There is a musty smell in the air, which mixes with something like savory umami, recently added by the Japanese to the long-standing European four: salty, sweet, bitter, and sour.

On the street, pickers squat next to baskets of a dozen species: *boletus, agaricus, amanita, russula,* and more. Some will be bought today and shipped off fresh to a large wholesale market. Matsutake have gained center stage, dealers amassing crate after crate as they pour in from distant villages. I think of some of my students who lived nearby. I taught at Southwest Forestry University in 1995, and one said he walked the last kilometer to his village by foot, as one part of the trail was too narrow for horses when it skirted a sheer cliff. Every morning, from places like this, people hunt mushrooms, and if they find fresh matsutake, they or their finds end up at this wholesale market: it is a pulsing miracle of activity.

Matsutake is not like a crop of wheat in Northeast China, grown on a thousand hectares, planted and harvested quickly by a few people driving massive tractors and combines and then dried, ground into flour, and stored for years. No, these mushrooms emerged in great diversity and profusion, in forests thick with human involvement, yet together with plants, animals, and fungi rich in their own life projects. Here, thousands of people are gathering them up, as underground mycelia, some intertwined but making their own relations, push up new pulses of sporulating bodies into the air.

—Reflections on fieldwork

This chapter moves us up to the mountains and into Yi villages to explore at the ground level how the Yi—members of one of the two ethnic groups that devoted themselves to the matsutake trade and were transformed by it—make their worlds in relation to this mushroom. As this book demonstrates, however, the matsutake economy in Yunnan (as elsewhere) is not entirely controlled and directed by human actions, by the activities of people, be they Yi, Tibetan, or anyone else. Rather, matsutake liveliness inflects the market that humans build around them from beginning to end. This chapter shows how matsutake liveliness shapes their participation in the economy in two main ways: as explorers and as performers. As described in the previous chapter, matsutake are explorers that found this part of Yunnan, and they are creative relationship builders that have figured out ways to live with these trees in this place. To do so, they had to reach out to some species and defend themselves against others, as they are an attractive source of food for many animals. Belowground, matsutake are constant performers; aboveground they make an annual display that is usually the only manifestation that most humans ever notice.

The Yi pay attention to these forms of matsutake liveliness and jostle their own daily and seasonal patterns of work in an effort to accommodate them. Villagers must attune themselves to the rhythms of matsutake's world-making, including its conditions of growth. At the same time, Yi are also deeply attuned to living with the plants they raise such as corn and buckwheat, and animals such as chickens, pigs, horses, and goats; they also recognize how they share the hunt for matsutake with a range of other species including insects. This chapter, then, begins with an ethnographic exploration of the Yi matsutake economy as seen from the perspective of Yi villagers. It then turns the lens around and takes a closer look at how matsutake's world-building capacities shape the many thousands of Yi-matsutake encounters that have taken place every year since the mid-1980s.

Hunting Matsutake Worlds
in the Mountains of Yunnan

As we boarded the bus in Kunming, Yang Xueqing and I were full of anticipation. Xueqing was a doctoral student at the Kunming Institute of Botany, and we were heading into the region where she was raised, the heartland of the province's Yi community and the center of Yunnan's matsutake habitat. We followed the newly widened highway to Chuxiong City, where we disembarked, bought snacks of chili peanuts, and transferred to a shuttle van, its seats torn out and shocks long past their prime. The now-bumpy road climbed ever higher, following the curves of the hills and leaving behind the wide valleys of rice paddy for the realm of corn in small plots, interspersed with roadside plantings of eucalyptus trees, harvested for their essential oils. Yunnan's landscapes, now a multicolored vibrant green from the monsoon rains, were always a source of wonder for me. Villages were dispersed, and within them, homes of adobe sunbaked brick built around wooden post-and-beam frames and roofed with dull-red clay tiles were compact. Homes of the wealthier families often sat behind walls with more ornate tiles of yellow and green; and starting in the late 1990s, people built concrete homes, many sporting green- or blue-tinted windows, a popular urban style at the time.

In the middle of the day, few people were out and visible, but we saw several farmers wearing dark-blue polyester sport coats and old olive-drab army shoes out hoeing their corn on the steep slopes. One farmer rested in the shade of a two-story drying shed built of rammed earth, which stood at the edge of the fields looking like an impromptu military tower. It was built for drying tobacco leaves, a new cash crop that had made its way up from the warm valleys below and into the hills. Yet in the hills with colder nights, tobacco lost its vigor and didn't grow well; eventually farmers left it for the lowlanders to grow. Another farmer was spreading fertilizer or pesticide, reaching a hand into his homemade satchel, the kind often made from the woven-plastic sacks used for commercial pig food, and gently tossed an arc of pellets and dust. As our van climbed higher, fields of green potato stalks emerged in striking contrast to the red tones of the clay soil.

Looking out the window while climbing up the hills, I reflected on this region's history. We were located just off some of the routes used by the tea and horse caravan that traveled from southern Yunnan and Thailand into Tibetan and Indian territory. Nanhua was above the wide valleys that contain rich soils and abundant water, perfect for growing rice. Much of this bottomland was claimed by Han settlers, many of whom came centuries ago as imperial soldiers and farmed with state support. These settlers often used an overarching term to describe the diverse surrounding groups: the "Yi," a general term meaning barbarians, "uncooked people," those beyond imperial control.[1] Only in the 1950s did Beijing conceive of China's rich ethnic diversity in terms of "ethnic minority" groups. After the first ethnic survey reported more than four hundred different groups in Yunnan alone, the reply from Beijing was along the lines of "too much diversity; simplify." The anthropologists went back to work. The plan was that each official group name was supposed to designate one specific ethnic group with one culture and language. Yet nonetheless, the term "Yi" was recycled to encompass a wide range of groups that called themselves different names and spoke at least six mutually unintelligible languages.[2] Thus, this was unlike some officially recognized ethnic groups whose membership coincided with their own self-description; Yi was again used as an umbrella term. Given that, I don't want to imply that there is a single Yi way of being or thinking, and so the statements I make refer almost exclusively to the time that I, sometimes with Yang Xueqing, Luo Wenhong, or Anna Tsing, spent with Nanhua Yi in Yunnan in 2008, 2010, and 2012.

In 2001, many of my professional friends in Kunming (who were mainly Han settlers who had arrived in the past fifty years) described Yi as "relatively backward" (*luohou* or *bijiao luohou*), poor farmers eking out a meager existence in the mountains. Relatively few were aware that the Yi had not always occupied this position.[3] We drove past hillsides where we could spot intricate and expensive tombstones carved a century ago, revealing that wealth had flowed through this region before. When I asked about this, one old man said he remembered some of the stones being erected in the 1940s, when he was a little boy. Almost in a whisper, he said the stones were likely paid for with opium money, a crop eliminated by the government in the late 1950s after nearly a century of providing massive flows of silver into these hills.

The next major cash crop was hemp, grown not as a drug but as a strong fiber demanded by the state until the 1980s. As I explain below, it brought some brief prosperity, but nothing like opium. One old man said that even though money was relatively scarce under Mao Zedong's rule, some people gained cash by growing hemp and processing its tough-wearing fibers to produce the ubiquitous sacks of rural China, used to carry grain or fertilizer or pig food.

He described making hemp as painful and arduous labor, mainly undertaken by women in cold mountain streams that numb one's hands; the work was so demanding that the main expression in China for "a lot of trouble" is literally to "make hemp" (*mafan*). By the 1980s, however, the state's factories started making sacks from plastic, not hemp; communes were dismantled, and most Yi had little to sell.[4] They became poor again. Later, a development project that encouraged farmers to grow tobacco seemed promising at first but did not work out well in this upland area, after a series of early fall frosts resulted in big crop losses. Tobacco, grown by Native Americans for millennia in warmer climes, was not always able to ripen in these hills.

As the state moved from hemp to plastic, and as agricultural subsidies began to diminish and prices for fertilizers and daily expenses began to climb, mushrooms came to the rescue again, first in terms of daily subsistence. As recounted to me, in the early 1960s, famine stalked the land throughout China, and many people died. Here, as granaries emptied of rice and corn, Yi started eating more and more things from the forest, including the inner bark of trees. A long-standing interest in mushrooms expanded under great hardships. New species were found to be edible, and others were found, tragically, to be poisonous. These were scary times as people experimented.

By the mid-1960s, the Yi here had adequate food again, and a number of mushrooms had become part of their diet, but matsutake were not especially valued. Children could make some spare change selling them. In fact, one old woman later told me, "We would add them to the soup pot. It was like adding tofu. But actually, tofu would have been better

FIGURE 5.1. Li Bo demonstrating how matsutake were pierced on a
stick. Photo by Michael J. Hathaway.

than matsutake." In the informal markets, they might sell a handful of
mushrooms pierced one by one with a thin tree branch, like a shish
kabob. Li Bo, a heavy-set Yi musician who had taken to matsutake hunting and was quite good at it, showed me how this was done (fig. 5.1). He
laughed at the idea of cheap matsutake and kind of winced when he
showed me how the mushrooms were pierced, saying how the stick

really hurt the mushrooms: "These days, no one would do *that* to a matsutake!" Picking matsutake changed so quickly, from something for children to do in their spare time, to becoming a family activity and a key part of their livelihood. Since the late 1980s, matsutake has become the area's main economic driver, replacing hemp and opium.

En route to matsutake country, we disembarked at Nanhua, a small town built along the one paved road of the region some people now call the "Kingdom of Mushrooms." I had already talked with Brian Robinson, a PhD student at the University of Wisconsin, Madison, who had come to Nanhua to study how villagers devised systems to divide up rights to natural resources, such as matsutake. Brian had told me that the town's matsutake king was Li Laoban (Boss Li), so I asked around among the shopkeepers, who pointed us toward his compound at the end of the road. Li Laoban was the designated representative of Nanhua, at the heart of some of Yunnan's most abundant matsutake lands. He became a spokesperson for Yi matsutake worlds. He had a tidy and well-swept concrete compound with a large guest room where he entertained visitors. He had taken care of many outsiders, mainly from Yunnan, who came to see how Nanhua created its booming matsutake economy. Wearing his signature white dress shirt and full of energy, after brief introductions, he asked if we wanted to see the demonstration forest, and of course we were thrilled.

Walking through a maze of small alleys, we passed active construction of new homes and shops, and then we came to the edge of town, which had seemingly endless forest behind it. There was a large sign and an extensive boardwalk through a large patch of forest surrounded by a wooden fence (a rare phenomenon, except in all but the most popular of tourist sites). He explained that the boardwalk was designed for maintaining the forest, to avoid soil compaction, and because visitors often wore expensive leather shoes or high heels and would not want to walk on muddy trails. My PhD adviser, Erik Mueggler, had spent several years with Yi communities in this area in the 1990s and had told me how they would select one family (gifted with wealth and loquacious speakers) as the *ts'ici* (a Yi word) who would generously host powerful visitors, whether warlords before the Mao era or today's officials. Although many warlords were themselves Yi, officials were almost always Mandarin-speaking Han. For the ts'ici, such a position was both an honor and a duty, for such outsiders could bring fortune or peril, depending on how they were treated. To assist the ts'ici, the surrounding families contributed hosting expenses, as visitors typically drank more and better alcohol than most villagers could afford, and they dined on chicken and pork in quantities greater than most locals would ever do. I imagined that Li Laoban was a modern-day ts'ici, welcoming important visitors and the occasional social scientist into his midst.

With pride, he rattled off the positions of the delegations he had hosted over the past few months. He explained that with the new national environmental law, which had followed the massive flood of the nearby Yangtze River, officials were scrambling to find new economic goods. Nanhua was lucky, was blessed, he said: it was a land rich in matsutake mushrooms, and his fellow villagers had found ways that many people could make money, not just himself and a few other laoban. He mused about the future:

> Someday, matsutake might disappear from the forest. Laws might put an end to the trade, or Japan might stop wanting the mushrooms from China—the Japanese might say they are too polluted. We are trying other ways to make money, but it is harder work and takes a long time to profit—planting walnut trees [to sell the nuts], creating Yi tourism. Some people went into mining, but it was dangerous. There are laws that prohibit that now. Most people just do a little bit on the side: work as a driver, raise some goats and sell goat cheese or meat, raise a few pigs or chickens and sell them to the butchers—

there aren't many ways to make money. Matsutake is the best at providing for everyone [nodding his head in agreement with himself].
Why don't you walk around town, meet my brothers and some of the
other dealers? Maybe next week my driver can take you up on his
buying trips. You need to talk with the farmers too.

We left Li Laoban's place, heady after such a warm welcome from such a
well-connected and knowledgeable person. Over the next month, we
met with a wide range of people, taking up his offer to accompany one
of his drivers into the hills to see the trade firsthand, and we walked with
farmers into the forest to hunt with them and see the land as they did. It
turned out that matsutake was just one mushroom out of dozens that
people picked: as summer turned to fall, mushrooms were everywhere;
vans packed with bags of dried mushrooms skidded around muddy corners, entrepreneurs stood with their metal scales at street corners waiting
for individuals to bring their daily haul. Groups squatted around massive
piles of fresh mushrooms—brushing off pine needles and debris and
dumping them unceremoniously onto wooden frames stretched with
wire netting, preparing them for a massive drier made of roughly built
pine but with ornate metal corners, looking new and also old with its
smoke-stained wood. At the same time, we couldn't help but notice that
matsutake were only rarely treated that casually; instead, they were usually moved quickly, sorted into piles one by one, and gently tipped from
one basket into another. Most dealers handled them delicately, especially
when the price was high, and hunters would take the nicest specimens
and swaddle them in leaves like a baby. Unlike the majority of other
mushrooms destined for China or Europe, which were dried (fig. 5.2),
matsutake were left unblemished—because the Japanese had high demands for fresh matsutake.

For the next month, Li Laoban remained an important guide for us
in understanding the rise of the Yi matsutake economy—how it had
changed people's lives, where and how they went to pick matsutake, and
the pace at which they moved to syncopate and align themselves with
the tempo of rains, mushrooms, and insects. After traveling with some
of his buyers, we would sometimes return to his large compound, which

FIGURE 5.2. One of the relatively few female entrepreneurs (on the left), who deals mainly in dried mushrooms, logging the date, weight, and price of these dried mushrooms. The young woman on the right has just sold these bags. Photo by Michael J. Hathaway.

at night was often bustling with people coming through, playing festive drinking games in a screened-off dining room. Li Laoban was typically a pensive person but could get excited when talking about his projects for the community. One evening he discussed his upbringing, a topic he rarely raised, and he remarked on the massive shifts that had taken place over the course of his life. It all began when I pointed to a plaque that Li Laoban had been awarded, celebrating his household as a "10,000-Yuan family" (*Wan Yuan Hu*). He chuckled, saying that it was given to him years ago, in the 1990s, when that was a lot of money to earn, but that these days it's really nothing—everything is so expensive now. Nonetheless, after local government officials gave him the plaque, he said it represented a big sigh of relief, for he could tell then that the winds of fortune were notably shifting, and that now it was good to be an entrepreneur, a boss of others.

He said that, in contrast, when he was young, in the days under Mao, communes were still in place, and many people had been punished for being seen as too entrepreneurial, as possessing, some said, a "capitalist tail," an allusion to their not-quite-human and indeed monstrous nature. Such language was part of the everyday discourse of the Mao era, which lasted twenty-seven years, from 1949 to 1976, the year of Mao's death. Three years later, China began another rapid and ambitious shift moving toward a market-led economy. Even so, they were still shaped by nearly three decades of socialist laws, policy, and sensibilities: the past was not so easily forgotten. There were still lingering feelings of resentment toward the rich and suspicions that they earn their wealth through the exploitation of others.

In part to respond to such accusations, Li Laoban was quick to point out, as many other laoban do, that he carries out his work at great risk. Matsutake is an incredibly volatile commodity; the price shoots up and then crashes down quickly seemingly more than any other good. As he said, "One day the price is high; the next day the price is low. I can't predict it!" No one really knows why the prices fluctuate so much, and the mushroom itself demands that he sell it quickly because the mushroom is delicate, and decay is rapid. He only has a day or two at most to sell off his collection, so if he buys a large quantity of good-quality mushrooms (fig. 5.3), they can quickly deteriorate and become worthless. One of his most trusted buyers, a man named Wen Bing, however, invoked a gentle challenge to his boss's discourse of risk: "With matsutake, you are pretty safe. You could only lose a day or two's worth of sales. Other laoban do business with a lot of goods; they can have a full warehouse of products. It can burn down, and they don't have insurance, or they won't pay." Li Laoban looks at Wen Bing and shrugs with a slight smile, and I'm thinking that this is a special relationship, because all of Li Laoban's other buyers seem to always show him obliging obedience, not this type of respectful cheekiness.

Nonetheless, for Li Laoban and others, the pressure is always on, and there is a sense of urgency to finish deals and move the mushrooms out to the export facilities in Kunming where they are tested and sent to Japan. His small crew of workers drives up the winding dirt roads into the mountains. They pull over to meet groups of men who combine the daily harvest into several large baskets so they can size up the offering

FIGURE 5.3. Three men discussing the final price for three crates of matsutake (on the left), while the buyer awaits. Photo by Michael J. Hathaway.

and propose a bulk payment. Li Laoban's buyers come to village stores where shopkeepers often buy just a handful of mushrooms at a time but nonetheless can accumulate a large number every morning. Back at his home, Li Laoban is also a retail buyer. He remains alert to new sellers coming in, and he is related to many of them.

They tend to treat him with both the familiarity of kin and also a kind of deference offered to bosses, who possess much discretion about how much to pay pickers for each basket of matsutake. Grading mushrooms is as much an art as it is a science, and it is typically difficult for pickers to find out the day's local price, except from friends who have already sold their daily harvest. Although the weight of the mushrooms on the scale is not debatable, the placing of each mushroom into one of five or six categories is a decision. As well, the price is an act of judgment. It's not that each mushroom is priced individually but that the whole pile is sorted.

The formula is neither scientific nor predictable. Prices can go up and down within a single day. Whatever the offer is, it is largely accepted by sellers, because farmers have few other dealers to offer their goods to. The other main contender is Mr. Wang, a half mile down the hill. In one conversation I overheard, two sellers gently debated what to do about this, with the younger man saying it was best to find a higher price. The older man, his uncle, suggested it was better to establish an alliance with one dealer, with some degree of obligation and responsibility, rather than playing each off against the other every day. His nephew didn't concur, so he drew a breath and explained that there are times when you may need the ties that loyalty bring—when you ask the boss for a loan or need some help getting a permit from the government, for example. Ultimately, too, there are hundreds of sellers and only a few dealers, so it's not easy to find other dealers to bargain with, and thus, the dealers usually hold the upper hand.

Li Laoban acknowledged this double sense of good fortune. First, he said, they lived in this place where matsutake were abundant; they were a gift from the forest. Picking matsutake was not like growing rice, which was a hard-earned crop requiring massive amounts of embedded labor to make rice paddies, reroute water across the landscape, and maintain a system that needs constant upkeep. Second, the Japanese had found out that matsutake were here, wanted them badly, and were willing and able to pay a high price. He found it puzzling that China was on the winning side of World War II and remained poor, while Japan lost the war and became rich. He sensed that this coincidence—their local abundance of matsutake and Japan's desire for it—was probably fleeting. They had to make the best of this situation while they could, before it changed. I heard similar musings elsewhere in China, the sense that winds could quickly shift, and that one had to be poised to make a major life change if conditions appeared fortuitous.[5]

Li Laoban reflected on how this mushroom economy was special in supporting the Yi language. In most other realms of life, such as education and business, fluency in Mandarin was increasingly becoming a prerequisite for success. In Nanhua, there are few nine-to-five jobs. Many small businesses, such as the shops selling pharmaceuticals, require workers to speak Mandarin, and most native Yi speakers are expected to be bilingual, although Han workers can get away without learning Yi. Engaging in the matsutake market represents one of the few realms in which both workers and bosses can make money while they converse in the Yi language. The high-level dealers all speak Mandarin well, but a number of individual sellers are far more comfortable in Yi than Mandarin. All day long, walking through streets thick with mushrooms, one hears Yi spoken among those gathered around a dealer's scale or on their cell phones, making the matsutake economy a shared linguistic community.

Some outside bosses have come here to buy matsutake, but one of the other main dealers in town, Zhou Laoban, said that they now come only rarely, and when they do, almost no one will sell to them. He smiles as he tells me this because it displays his sellers' loyalty to him. He says his sellers have a good relationship with him because he tries to give them good prices. He will sometimes err on the generous side for the older widows who do not have much local family support, such as filial sons or daughters. He has room to decide as he sees fit.

Sometimes, outside bosses come with plans that don't have to do with buying mushrooms. Occasionally, they work out, such as a sprawling recycling operation created alongside the road. Other bosses hatch schemes that turn out to be fly-by-night ventures: one day while up in the woods looking for matsutake, a villager pointed out hundreds of pine trees with a V-shaped cut on them. "What was this?" I asked. "Oh, two years ago an outsider boss offered to buy a thousand liters of pine pitch to sell in the city. We tapped all these trees. The boss never came back." I heard about a number of cases like this, where outsiders proposed projects that never came to fruition, so villagers are often skeptical of new plans. The matsutake economy, however, has remained. The price goes down and up, and because incomes are so low in this area, matsutake continue to be very important.

The Fruits of Matsutake: "We Are Building a Yi Economy"

One afternoon when Yang Xueqing and I stopped in to see how Li Laoban operated his sales office, he invited us to dinner. Li Laoban expanded on last week's point that matsutake picking was important because people could make a decent living while speaking their own

FIGURE 5.4. Yi maidens dancing in spotless regalia at a large festival.
Photo by Imaginechina/Alamy.

language. More broadly, he said, he and other local laoban were using matsutake funds to help build a Yi economy. We walked down the street, and he pointed out an example, a relatively new restaurant selling Yi food, which turned out to be one of his investments. The restaurant offers a number of Yi specialties that don't commonly appear in Han-run establishments, such as goat meat, a kind of fried goat-milk cheese (called *rubing*), and some wickedly strong and locally distilled liquor, as well as some famous brands of maotai. On the walls, calendars from local companies display Yi scenes, such as Yi dance festivals, with Yi maidens in spotless and elaborate regalia (fig. 5.4).

These have become places where he can entertain guests and show-case his own cuisine. He says that he sometimes meets other laoban from out of town, often Han, who have come to search for business opportunities. He likes to take them to eat at his restaurants, have them eat Yi food, and drink Yi liquor. "Many of them don't know our food; they aren't used to it," he says. He has been to Kunming, the provincial capital, many times and says that "in Kunming, Yi restaurants are rare;

maybe one or two will open and then close. I have two Yi restaurants in this little town." I can't recall noticing a Yi restaurant when living in Kunming, even though other forms of ethnically marked food, such as Dai cuisine, have been quite popular for decades, and since the early 2000s there has been a greater quest by urban Han to encounter forms of "difference," culinary and otherwise.[6]

Like the Yi restaurants, Yi cultural centers are mainly enabled by matsutake wealth,[7] following a pattern of matsutake money being invested in other ventures—low-end hotels, restaurants, and ethnic-based festivals. At these smaller events, there might be certain iconic foods with appeal across ethnic groups, such as buckwheat cakes dipped in honey. By itself, buckwheat, which is called "bitter wheat" (*kuchao*) in Mandarin, might be too unappealing for those unaccustomed to its taste, but with honey it is a crowd pleaser.

Nanhua-based markets exist mainly for local people, not tourists from the big cities. In Nanhua, the town market takes place every Sunday, and many people, especially women, come dressed in their finest Yi clothing. Yi villages are widely dispersed, and some visitors to the markets tell me they spend several hours walking along dirt paths, from hamlets in the surrounding mountains. Many come with small bags of their own goods to trade, such as baskets full of piglets, handfuls of chili peppers or squash, and bunches of wild greens, and to buy clothing or packaged food such noodles, soy sauce, sugar, and salt. This kind of trading network was established long ago and in some places has been in existence for centuries.

This spatial arrangement of local markets, it turned out, coincided well with the matsutake economy. In this region, groups of people with few economic opportunities, who knew the forest and landscape well and who were already attuned to mushrooms, were drawn to the daily hunt to earn precious cash. This long-existing system, in which many are petty traders already, is one of the keys to Yunnanese dominance in the global matsutake trade. It is not just the maintenance of ancient tradition, for during the Mao era, such trade networks were diminished as state officials pushed rural citizens out of earning money through trading and crafts (such as making paper or cloth) and toward growing grain. Thus, from 1949 to 1979, a range of economic opportunities disappeared; when new ones emerged in the 1980s, they were seized upon.

Within the past few decades, during the several-months-long matsutake season in the late summer and early fall, things happen differently, and the whole town bustles, moving to the beat created by matsutake's tempo. For Yi in the most productive areas, like Nanhua, even for those not directly involved in hunting and selling matsutake, their own lives are often deeply affected by matsutake wealth. Restaurants, for example, tend to do a brisk business during the fall hunt as some residents are too busy and tired to cook but feel temporarily flush with cash. No one can walk down Nanhua's main street in the fall without stepping over mushrooms, and many of the new vehicles on the streets are owned by people who became mushroom dealers. There is no real tourist draw here, but during the mushroom season it is a lively place with lots of commerce. The only steady stream of outsiders that come here are officials who want to see Li Laoban's fenced-in matsutake patch, and to hear how he helped make Nanhua a kind of model for "getting rich" from matsutake.

In the urban zone of Chuxiong City, about an hour away by bus, residents use mushroom abundance as a tourist draw. Mushroom-based festivals have now become an important part of the seasonal round. The Wild Mushroom Gourmet Festival, for example, began in 2003. It featured "ethnic cultural performances" (which generally meant Yi singing and dancing), a wild mushroom industry forum, a trade fair, and a competition for ten winners of the "King Wild Mushroom" contest. Within less than a decade, this festival was attracting more than ten thousand visitors, mainly Han people from Kunming. Previously, Chuxiong had been only a rest stop for buses from Kunming on their way to well-established tourist sites like Dali and Lijiang. As events get larger, officials always get involved, in part because they are always looking for new development projects. And government involvement is not unwanted, especially for larger events, for only officials can bring powerful connections and can initiate projects on a scale that is possible only with state assistance. Abundant and valuable mushrooms became a way for this region to be more successful in the increasing efforts of towns to compete with each other for attention, sometimes to foster tourism, but also to attract the praise of officials and to solidify connections with them.

Still, Yi have a long way to go to successfully establish themselves as equals to the Han as there is a well-entrenched sense of Han cultural superiority and pride. This extends even to some of the recent Han migrants who are relatively poor and engage in low-status work. I recall meeting a Han laborer who was repairing a sidewalk. Even though his clothing was torn and stained, and his face bore a history of many years working outside in all kinds of weather, he opined that these people needed Han to come and show them how to build a city. Looking around to make sure no one could hear him, and almost whispering, he said, "The Han people understand science; the Yi do not." He saw me, as a Western foreigner, as someone, like him, who understood science in a way that he thought Yi did not. He had a strong sense of belonging to a group with a clear monopoly on national political power and cultural prestige, even though he experienced low social status in his everyday life.

In Nanhua, I heard a number of people describe the link between matsutake wealth and cultural pride. One midlevel dealer, squatting on the sidewalk with his metal scale and large basket of already purchased porcini, said, "We have been poor. We live in a Han economy [but] we are building a Yi economy." I was interested that he, too, used the term "Yi economy," which I had heard earlier. He said, "Mushrooms are king here. Matsutake (Yi: *mene*) are the biggest king."

His friend explained that it was hard to get good jobs, like middle- or high-level positions in government, which mainly went to high-school- or college-educated Han who could speak fluent Mandarin and who had good relations with other Han in positions of influence. Relatively few Yi in rural areas finished high school unless they had close relatives living in town where the high schools were located, as students often had to board overnight at the school or in town, and the fees associated with high school were constantly increasing. Becoming a low- to medium-level matsutake dealer, on the other hand, had become something in the realm of possibility for Yi youth, as it did not require a high level of formal education or strong connections in the Han-dominated world. For many villagers in outlying areas, it was often challenging for parents to send their kids to high school for academic and financial reasons. Some big-town high schools were academically competitive, demanding a high test score. Financially, high school could be burdensome for rural families, whose children had to move to a sizeable town, and unless they had a generous relative who lived there, the added costs for room and board were beyond their means. One of the few ways to finance this leap was to become a lower-level matsutake dealer, collecting from peers and selling large baskets full of matsutake in town.

Thus, the matsutake network was capillary, as many people participated as pickers and dealers. Many lower-level dealers tried to leverage their relatives and friends to sell to them. The higher-level dealers were generally not bothered by these lower-level players, so long as their own supplies of mushrooms remained intact. As prices for buying and selling were not substantially different, dealers made money mainly through buying and selling in vast quantities, earning just a bit of profit per every kilogram bought and sold.

Multispecies World-Making: Trees, Insects, Yi, and Others

Above, I have provided a sense of how the Yi people have engaged the matsutake mushroom in their own world-making, which at this juncture is focused on building up Yi culture, status, and identity. Thus far, humans have been at the center of my narrative, but now I turn the lens around and focus on how matsutake itself engages in its own world-building activities, with trees, insects, and Yi people. From the above, it seems obvious that Yi, as well as the hundreds of thousands of other pickers and dealers that create Yunnan's matsutake economy, are primarily motivated by the wealth they can earn from selling matsutake as commodities to Japan.[8] And it is easy to understand matsutake as a mere commodity, a preexisting "natural resource" that was pulled by humans into human-controlled economies. But when one displaces an understanding of humans at the center of the universe, other kinds of dynamics become visible. We begin to see how matsutake themselves are lively agents that are engaged in making their own worlds, and that in some places and times, humans connect with these matsutake worlds and at other times not.

One way I have begun to conceptualize matsutake in a more active way is to think of them as having three qualities. I started to appreciate that many species that have appeared on Earth are now extinct, and the very fact that matsutake even exist in Yunnan, or anywhere else in the world, is due to their actions as explorers, relationship makers, and performers. Such active terms are typically reserved for humans, but I think it is reasonable to imagine that fungi also possess and avail themselves of these qualities.

Let me briefly rehearse a few points from the previous three chapters, which foreground matsutake's active engagement with the world and world-making. As *explorers*, they had to arrive in Yunnan, either by spore or by mycelium—the former possibility requiring successfully launching their spores into space to catch the winds or hitching a ride on animals that carried them; the latter possibility requiring traveling slowly by mycelia. In order to survive in new places, almost immediately the hyphae had to be expert *relationship builders* and adaptive in their actions. They needed to create the relationships quickly, because they would not have had long after their spores germinated to make connections with trees. It is probable that the trees in Southwest China were different from (but similar enough to) those found in the places these spores had traveled from. Thus, matsutake learned how to form intimate semiotic and material relationships with different kinds of trees and roots, sharing and communicating by chemical compounds that are exchanged through the soil or the air.

In what ways are matsutake *performers*? The perspective of this book is to move away from the commonsense understanding of performance as a human-only activity, and as something consciously "acted," apart from everyday life, on a stage for others to see. Instead, my understanding of the term *performer* is based on the social scientific concept that actions can be seen as a kind of performance. Scholars like Judith Butler, for example, argue that people live as gendered subjects in a million everyday performances that include how people dress, speak, walk, and talk.[9] To be rendered as intelligibly gendered subjects requires meeting certain audience expectations. Extending this idea to a mushroom, we might see how matsutake, like all forms of life, are always performing in relationship to both living and not-living others. Their performances include all the many acts required to stay alive: finding new sources of food and water; mediating attacks from bacteria, insects, and others; exchanging nutrients with trees; and growing and dying. As well, to carry on the next generation, the fungi actively lure in insects, mammals, or birds to carry their spores to new realms or propel them powerfully into swirling currents of air.

Most of these actions are largely imperceptible to human umwelten because mushrooms often move too slowly for humans to notice. Time-lapse videos reveal mushrooms as active and moving subjects that we do not see as in motion with our limited perception and patience: popping into the air, unfurling their cap, and ejecting their spores. More difficult to see in underground terrain, slithering hyphae explore, drink up drops of water, and drill into grains of sand for their nutritious minerals.

These three main arenas of matsutake world-making capacities compel the Yi (and others) to respond to matsutake as it grows in the forest. Yet even after it is picked, its own liveliness continues to shape human responses. One might imagine that as soon as a matsutake is severed from its network of underground mycelia, it dies and becomes a passive object under human regimes of mastery and control. But the mushroom is still alive, just weakening over time. It continues inhaling oxygen and exhaling carbon dioxide long after it is detached from the ground. That is part of the reason why mushrooms are not supposed to be stored in plastic bags: they might suffocate for lack of oxygen.

As well, within the body of a matsutake, a whole world persists and even grows, taking advantage of the mushroom's weakening defenses. Almost always, one or more species of insects are eating the mushroom from the inside out. As insects have penetrated matsutake's protective skin, bacteria find easy passage through these tunnels and multiply thanks to the mushroom's weakening immune system. The mushroom loses its rigidity and can even get moldy, evidence that other fungi have joined the fray.

All those people who have gathered and sold matsutake on its route to Japan have observed the way insects and bacteria can make short work of a fresh mushroom, and thus dealers work as fast as they can to outpace their nonhuman competitors. Previously, I had imagined the decay of vegetables or fungi as a passive or inevitable phenomenon, mainly because they were cut from their feeding structures. This is because I had failed to see how it was precisely the liveliness of bacteria (especially their actions of breaking down animal and fungal bodies) that was creating this process.[10] Thus, it is not only the Japanese desire for fresh matsutake but also some of the qualities of matsutake's own world-making—such as the cluster of insects and bacteria that seek out these mushrooms to make worlds together—that compel the contours of Yi actions that make up this fast-paced, ice-pack-cooled, and highly attentive trade in these precious fungi.

Recognizing Insects as Fellow Matsutake Hunters and World-Makers

In talking with Yi villagers, I encountered a way of thinking that took insects more seriously as agentive beings than I ever had before. Squatting down on a small patch of flat ground underneath a clump of scraggly looking pine trees, Li Mingwo reached under an unlikely bush and pulled out a surprisingly nice-looking and fat matsutake, calling out: "Hah, hai meiyou zhaole!" (Hah, [another one that] hadn't yet been found!)

She said that even though the fall had been drier than usual, this patch of ground, with a small dip in the middle, could catch the rainfall. And with soil that was well packed from thousands of footsteps, it created a better spot for mushroom growth than the lands all around. She brushed some leaves and sticks from the top of this lovely specimen, which had probably started hitting the air a week ago. Splitting it in half, she nodded at it, showing me a dozen small tunnels created by insects. "When we look for mushrooms, *they* are also looking for mushrooms. The insects are better at it than us. What do we do?" I asked about how she and others had dealt with the insects over the past few decades, and she replied, "Let's ask Pa," as we headed back to her house.

I had met her father, Lao Wu, the previous week; he'd been sitting in his usual spot, a once thickly padded chair now worn thin, sipping rice moonshine while half paying attention to the TV in the corner. He leaned back in the chair and told me that in the mid-to-late 1980s, at around the same time that the Japanese had told them about matsutake, he and his fellow farmers, after years of receiving visits from agricultural extension agents, were just becoming convinced of the value of pesticides for killing insect pests. They started using it on their fields, and sometimes it really worked. Some years later, he recounted, some of his friends had said, "Why not use the spray to protect the mushrooms?" Matsutake was the most valuable thing they could sell, as they mostly raised grain and vegetables for their own consumption. It seemed like a good use of the chemicals; they could use a backpack tank with a nozzle and give each mushroom just a quick spray. But then in 2002, they heard the horrible news: the Japanese had shut down the Yunnan matsutake market after pesticides were detected on them. The demand for Lao Wu's mushrooms, which had been building up slowly, suddenly came to a grinding halt that led to a kind of panic. As he said,

> This was sometime in the early 2000s. The price had gone up and down for about ten years, but it [matsutake] was always worth more than other crops. Some people anticipated they would make a lot of money selling matsutake; they had taken out loans for a new motorcycle, but the market was gone. They had to look for money to pay back the loans. It caused a lot of trouble. We didn't know what would happen.

After this, many officials began criticizing the villagers, saying they were too backward, spraying pesticides on the mushrooms. He was both surprised and not surprised by this comment. "For years they told us to use pesticides. This showed we were modern and no longer backward. But, then, when someone used them on mushrooms, the officials got mad and said we were backward. Before, no one seemed to care about pesticides on the crops. Mushrooms were different." Officials warned them that the trade might be stopped forever because of the pesticide spray; but sales resumed. This led Lao Wu and others to work with Su Kaimei, a regional scientist, to find ways to protect their matsutake without spraying chemicals. Li Mingwo and her fellow villagers already understood that the insects were "better at finding matsutake" than they themselves were. They saw insects as full of ability, as active subjects, with some senses that were more powerful than human senses, and these views shaped their experiments with them. They said that insects, like humans, are individuals, and that they hunt matsutake with varying levels of skill. I had viewed insects as operating according to fixed instincts, often as pests to be dismissed or eliminated. In contrast, Li Mingwo and others saw insects as worthy adversaries, as fellow competitors, full of ability and cunning. Insects were hunters—indeed, hunters of mushrooms, like them. There were others, too, for mammals and birds were also actively hunting for matsutake.

Together with scientist Su Kaimei, the Yi began conducting forest-based experiments to test the insects' abilities *as subjects* and to see how they would behave in novel situations. Su Kaimei had been working in a lab in Kunming but wanted to help villagers address their particular challenges. Following the incident when Japanese importers detected pesticides on Yunnan's matsutake, she went to Chuxiong to collaborate with Yi mushroom hunters; they wanted to know what kind of world-making these insects do. They created experiments to see how these insects hunted—was it mainly by sight or smell? They devised tests to block the insects' ability to see matsutake (by piling up leaves around them to make them invisible from above, and later, by disguising them under a pile of sand). Next, they tried to block the smell of matsutake by covering them in clear plastic wrap (like Saran wrap) to see which insects were stymied by the lack of smell.[11] Observing these experiments, they asked, will the insects learn over time, or will they perhaps learn from each other? How and when will insects learn that a pile of sand means there are matsutake inside? Will this image of a pile of sand garner a new meaning and become something they search for? Those who were engaging in these experiments saw insects as agentive beings that could learn from their own experiences and teach others, directly or indirectly. They asked, Will insects that hunt by sight notice the presence of insects that hunt by smell? The mushrooms might be out of sight, but not out of smell.

In their experiments, they ventured that rather than finding their prey directly, these insects might take hints from other species, in the same way that other species—including humans—do. Recall that candy cane plants become signs for human hunters, because they are better at detecting matsutake mycelia than people are. Yi mushroom hunters' interest in studying insects' sensory worlds and hunting strategies likely arose in part because of Yi cosmologies—influenced by shamanism, Buddhism, and Taoism—which see the world as full of sentient beings. In my own cosmological training, I was taught to follow a secular, science-informed notion that largely dismissed the idea that animals could carry out purposeful actions, especially those animals that are considered less "evolved" or "complex," like insects. As I was explaining Yi insect experiments to a peer in Vancouver, he shook his head with skepticism, stating, "but insects don't have brains." I wasn't even sure if that was true (and it turned out to be untrue), though it resonated with my own upbringing. Back then, I hadn't considered that organisms might actively notice and interpret the actions of other species that were not their direct predator or prey. I later realized, however, that there were lots of examples: deer notice the behavior of squirrels who sense wolves coming; birds listen for alarm calls of other species; coyotes look to the sky to watch vultures soaring, triangulating their flight to find a carcass below; and human fishermen search for flocks of seagulls to find schools of small fish. Previously, I would not have imagined insects as capable of such observations and interpretations, but as I started to piece together a world-making approach, I recognized that these questions were excellent examples of what kinds of research questions such an approach opens up.

Conclusion

What I learned from Yi hunters and their collaboration with Su Kaimei helped me question assumptions that I had carried about ranking animals according to their intelligence and fostered my interest in learning more about the "minor traditions" among Western scientists—including those like Uexküll who presumed that all beings actively perceive and interpret the world, and do so with senses that we have yet to imagine. Yi experiments, however, also showed me a potential weakness in understanding the umwelt as a fixed sensory capacity; the Yi mushroom hunters I talked with saw insects as actively learning and coming up with new ways to act in the world in ways that mirrored their own improving abilities to find matsutake. They learned to attune their bodies to the places on the landscape that harbored matsutake, to poke at the mound that may belie the growing button underneath.

These experiences also motivated my quest to extend Uexküll's interest in animal perception to fungi themselves, asking how they smelled the world, how they sent out their own forms of chemical communication, and what kinds of senses they might possess. I read what I could find by the few scientists exploring these questions, and, extending their knowledge, I stretched my imagination in new directions.

The time that I spent in Yi communities also inspired my curiosity about how the other ethnic group most involved in the matsutake trade—Tibetans—might be encountering the mushrooms and their fellow hunters: humans, insects, and others. I had learned about sweeping environmental laws from my friend Xue Hui, who I knew about through her grandfather Xue Jiru, a prominent botanist. We had met in 1995, and by 2006 she had completed a master's degree on Tibetan pastoralism in Yunnan.[12] Xue Hui told me that these new laws were impacting Tibetans' age-old relationship to yaks. Other friends introduced me to a young Tibetan man, Norbu, and together the two of us traveled to Tibetan areas to see how they were organizing their lives with matsutake.

CHAPTER SIX
Tibetan Entanglements with Plants, Animals, and Fungi

When the earth tilts, the rain comes, and the cold returns, matsutake burst forth as their hyphae furiously knit together beautiful little buttons that slowly unfurl their caps to shoot their spores into the wind, and even make their own fungal breeze to carry them off in search of new lands.

A young Tibetan woman exclaims, "Be sha!" (oak mushroom), the last sound that echoes through the pine forest before the little matsutake is pried up from the ground and sent hurrying on its way—from a satchel made of old fertilizer bags to a large basket carried with a tumpline, to a series of cooled Styrofoam cases with packs of cold antifreeze, to a large warehouse in Kunming, where they are sorted again, washed, and inspected for chemical contamination and iron slivers (added to make them heavier), to the warehouses of Tokyo or Kyoto stacked with crates full of a million matsutake, whose striking scent of spicy cinnamon—the same smell that has lured in the many millions of insects all over the world, whose own larval children are now tucked away in matsutake flesh and eating their way out—catches the breeze and a passerby catches the scent: "Ahhhhh . . . Matsutake, fall has begun!"

—Reflections on fieldwork

In 2010, during my first stint of fieldwork on the matsutake economy, I left Li Laoban and other Nanhua Yi, and I got into a shuttle van heading up the highway to territory populated by Tibetans. With the Yi,

Tibetans are the other major ethnic group in Yunnan that shares land with an abundance of matsutake, and like the Yi, many of them have decided to dedicate themselves to the hunt. I was eager to find out what was happening with the Tibetans' matsutake economy. I had heard some rumors and read a few short papers that suggested that their economy was quite different from what I had just seen and that the newfound wealth had created unprecedented opportunities, but also much debate and conflict between villages.[1] The money had also helped fuel a building spree of giant new homes. Township officials endorsed the mushroom trade after a dramatic series of federal environmental laws came down especially hard in the region and government coffers dried up. Tibetans, it seemed, were selling off their yak herds in an effort to create a new way of life. These rumors and statements made me wonder what was going on and what kind of entanglements were happening.

I had first come to Northwest Yunnan—near the borders of Burma, the Tibetan Autonomous Region, and Sichuan Province—fifteen years earlier, in 1995, when the matsutake trade was already quite lively, but I didn't notice it at the time, as its markets were hidden away. Now I was heading north to spend time with Tibetan hunters, living in their homes and attending the markets in villages and towns to see how people's lives were transforming as they entangled themselves with these mushrooms. What I realized, more deeply than with the Yi, was how these new entanglements were always in relation to existing ones—and in this place, the most central relationship that had long shaped daily life and a sense of identity was with the yak. In other words, Tibetan world-making before matsutake had been very much focused on the needs and qualities of their yaks. But as things started to shift in terms of the lure of the matsutake economy and new laws, some people began to sell off their yaks, and this reconfigured their lives. This chapter, then, explores how the world-making activities of Tibetans, yaks, barley, and matsutake worked out in relationship to each other as living beings—beings that were subject to changing political winds that, together, created changes for all residents of Northwest Yunnan.

About six hours after leaving the Yi area, we were now back in another island of matsutake habitat. We had gone up in elevation, and the temperatures cooled enough for the species of pines and oaks that matsutake partner with to grow and prosper. You know when you've reached matsutake territory, as farmers from surrounding villages head to the highway to patiently squat alongside the road with baskets of matsutake for sale.

Heading north another few hours, reaching the large town of Gyalthang (Tibetan for "royal plains"), also called Zhongdian (Mandarin for "middle outlying fields"), I arrive in what is solidly Tibetan territory. In 2014, the town was renamed Shangri-la—a term that first appeared in English in James Hilton's 1933 novel *Lost Horizon*, but which has since become synonymous with a fabled land of peace, abundance, and wonder—to enhance the tourist draw. Such efforts paid off: in the space of just fifteen years, from my first visit in 1995 to 2010, tourism had increased nearly one hundred fold, from sixty-six thousand to over six million. Many foreigners came, but domestic tourism was also skyrocketing, and the vast majority of visitors were urban Han Chinese professionals, a demographic with a growing interest in Tibetan Buddhism.[2] Their presence, and the infrastructures built for them, largely ended up fostering the matsutake economy, which also led to reconfiguring Han-Tibetan relations and allowing Yunnan's Tibetans to reconnect with their compatriots throughout the larger region beyond the province (fig. 6.1).

Although these Han tourists came in peace, Han and Tibetans have a long history of military conflicts, and the region has shifted back and forth between each group's political domination. Evidence of Tibetan influence is clear in this striking building—the Ganden Sumtseling (Tibetan) or Songzanlin (Mandarin) Monastery—which still dwarfs anything ever built in the region by the Qing dynasty or the People's Republic of China (PRC) (fig. 6.2). It was started in 1679, designed

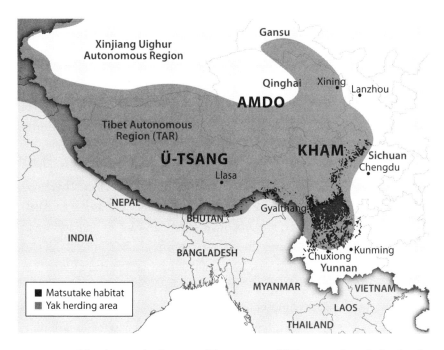

FIGURE 6.1. Map showing the three main Tibetan regions (Ü-Tsang, Amdo, and Kham) and the most productive areas for matsutake and yak herding. Map by Saki Murotani.

FIGURE 6.2. Ganden Sumtseling / Songzanlin Monastery in Gyalthang/Shangri-la. Photo by UlyssePixel/Shutterstock.

along the lines of Lhasa's Potala Palace (the political center of Tibetan monasticism) and was finished in a remarkably quick two-year period. Now, three and a half centuries later, it remains Yunnan's largest Tibetan Buddhist monastery.

Throughout the first half of the twentieth century, political control was uncertain as the Nationalists and Communists each vied for domination, and much of the area was run by local warlords or monasteries. In 1904, British colonial forces invaded Tibet from India, setting off a series of reactions that reverberated throughout the region. This invasion fostered antiforeign sentiments; soon, Tibetan lamas began to attack French priests and Tibetan converts to Catholicism (including in Northwest Yunnan). In turn, the Qing army worked harder to claim and govern this region, and the soldiers defeated the lamas and others.[3]

In 1949, the Communist victory that established the People's Republic of China was almost completely Han led. At first the new government worked cautiously to implement massive social changes in Tibetan-dominated areas. Just seven months after land reform (seizing land from monasteries and those labeled as "landlords") began in Gyalthang, in March 1957 Tibetans rebelled, and the military arrived quickly and put the rebellion down.[4] Then, within two years, a major Tibetan rebellion began in Lhasa and spread throughout the region, including to Gyalthang. This time, fighting was not just Tibetans versus Han; in Gyalthang, it seems, many lower-status Tibetans joined the militia of the CCP (Chinese Communist Party).[5] After much fighting, this rebellion was also defeated. As a result of the warfare, more Han moved to the area, and almost all lived in town; by 1964, according to the census, Han made up almost half of the town's residents and Tibetans about a quarter.[6] In the mainly rural prefecture, Tibetans remained the majority group at about 33 percent in 2010, with the next largest groups being the Lisu at 27 percent, Han at 18 percent, Naxi at 12 percent, and Yi at 4 percent; these figures evidence the overlapping ethnic diversity of these lands.[7]

I realized that Shangri-la was at the edge of Tibetan and Han frontiers, located more than four hundred miles from Kunming, five hundred miles from Lhasa, thirteen hundred miles from Beijing, and fourteen hundred miles from New Delhi. It is on the southern edge of a vast area

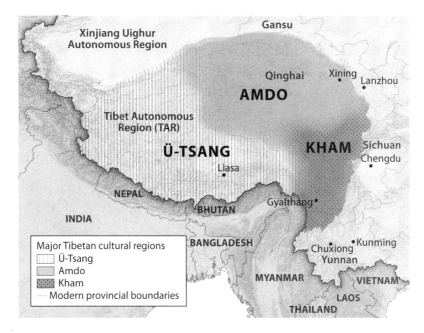

FIGURE 6.3. The three major Tibetan cultural areas and the modern provincial boundaries. Note that Tibetan presence extends beyond the Tibet Autonomous Region into Yunnan, Sichuan, Qinghai, Gansu, and the Xinjiang Uighur Autonomous Region. Map by Saki Murotani.

of Tibetan culture that extends far to the west to the Tibet Autonomous Region, to the north to Qinghai, and somewhat to the east to Sichuan Province. This realm is often divided into three districts: central Tibet (known as Ü-Tsang), with its political center in Lhasa; a northern area known as Amdo; and a southern area known as Kham (fig 6.3).

In the middle is the world's largest upland grasslands, an area about the size of Greenland that contains almost 80 percent of all land above four thousand meters. Although land above thirty-five hundred meters is beyond the limits of agriculture, it is the perfect place for yaks.

Gyalthang sits at the southern edge of Kham, one of its warmest areas with the lowest altitude, a place with more trees than pasture. When I first traveled here in 1995, I saw the Horse-Riding Festival, where nomadic Khampa rode horses with great agility and prowess, developed over many years of their taking yaks from pasture to pasture with the changing seasons. These men, sometimes known as the "Native American Cowboys" of China, are possessed of tall, powerful frames, their large felt hats reminiscent of cowboys'; and many of the Khampa women have long, straight black hair that recalls Native Americans, an image accentuated by their abundant silver and turquoise jewelry. The festival was a place of connection that brought together two groups of Tibetans who each hunted their own fungal treasure: the Khampa searched for the caterpillar fungus on the grasslands, and the Tibetans around Gyalthang hunted for matsutake in the forest.

People connected their pursuit of and trade in these fungi to their historic role as traders in the caravan of the Southern Silk Road, which for centuries brought bricks of fermented black tea from Southeast Asia into the Himalayan Mountains and to India. Yet during the Mao era (1949–76), Gyalthang residents went from living in a hub of activity along a major trade route to living in a place viewed by the new national government based in Beijing as relatively remote and isolated, on the edge of China. Under Mao, the Chinese government actively created and guarded national borders (with India and Burma) and drastically curtailed private trade, bringing this caravan trade to a grinding halt. Starting in the 1980s and the 1990s, locals became excited about returning to their role in a lively trade economy after a lull period of about three decades. As some people said, the matsutake economy helped them become international, *again*, and regain a stronger presence in the world.

A Detour through Interspecies World-Making

In this initial discussion above, I have offered a familiar story wherein humans are the main actors, the ones that count. The other animals and plants are mainly invoked as commodities (such as tea or horses) or sources of people's food (such as rice or barley). In the previous chapter, I stressed how Yi matsutake hunters and dealers responded to the actions of the mushroom itself, which altered the course of their lives. In this chapter, I stress how matsutake as an organism became entangled with Tibetans, who were already deeply engaged with yaks, barley, and trees. If we take an interest in how organisms make worlds in relation to each other's presence, then these accounts might look quite different. How, for instance, does the particular liveliness of yaks shape the lives of people who raise them? How does the specific liveliness of matsutake shape the lives of the people who pick them? How do we see these animals and mushrooms as active agents, making worlds together with a particular set of people? In Northwest Yunnan, I became acutely aware of how world-making is rarely just between two species but almost always involves forms of relationship with multiple kinds of liveliness all at once.

Like any people whose lives revolve around raising plants, animals, or fungi, Tibetans have been building and maintaining deep forms of attunement to particular organisms they look after. In this high-elevation landscape, Tibetan food and income have focused on raising one plant (barley) and one animal (the yak), and Tibetans have adjusted their lives to accommodate the needs of and threats to the barely and yaks. Before we turn to their recent entanglements with matsutake, we might step back to reflect on how the liveliness of this plant and this animal played an important role in expanding Tibetan territories. Looked at another way, in building a life centered on the flourishing of yak and barley, these relationships played a formative role in shaping the very being of being Tibetan. As the tenth Panchen Lama said, "No yak, no Tibetan people."[8]

Of course, such relationships between and among organisms were not solely determined by their own ways of living but were also inflected by political forces, their existence in a commodity economy, and then a capitalist economy, and so forth. This is not new; as mentioned above, these lands have been subject to overlapping and conflicting forms of rule and conquest for thousands of years. Here, I look at one major political force that shifted the relations between matsutake, trees, yaks, and people. On the other hand, such political and economic forces do not just operate *above* these organisms but work in relationship to them and their unique properties, actions, and relationships with others. In this case, a political shift pushed people away from cutting trees and herding yaks and moved them toward hunting matsutake.

Thinking back to my time in Nanhua, I realize that Yi communities, too, were entangled with a number of organisms, but these consisted of a broader range of plants and others such as corn and buckwheat and animals such as pigs, oxen, and goats. Among the Nanhua Yi, almost every family raised pigs but relatively few herded goats. Goats and pigs are highly adaptable animals that can live within a wide range of conditions, from low to high altitude, and both of them are famous for their ability to eat nearly anything. This incredible flexibility is part of pig world-making and goat world-making. Yaks, on the other hand, are highly specialized animals, suited to high-altitude mountains and only really capable of eating fresh and dried grass. Tibetans use yaks as their life center to enable them to survive and thrive in these high-altitude mountains, and they also devote themselves to the yaks' needs. These are forms of mutual world-making. As Melissa Johnson writes in her beautiful book *Becoming Creole: Nature and Race in Belize*, "human being is relational; people become who they are through their entanglement with each other and with agentive and powerful more-than-human entities."[9] This is important, because in more common understandings, humans are regarded as dictating the worlds of animals, but not being shaped by them in return. And we can see that in a deeper way, such relationalities are not limited to humans, of course. As we'll see, the yaks here also become who they are in relation to their bodies and personalities and through their daily relations with powerful and agentive human and nonhuman entities.

These entanglements or relationalities are always evolving, changing, and moving. Sometimes this happens quite slowly, and the relations are long enduring; the entanglement of Tibetans and yaks goes back millennia, but Tibetans' intense relationship with matsutake goes back only to the 1980s in this place. Organisms enter into entanglements in ways that are not always intentional, and the outcomes are rarely predictable, and yet their own specific liveliness deeply shapes the formations that emerge. Let's now look at some of these entanglements of barley and yaks, emphasizing the history in natural history.

Tibetan and Barley World-Making

Around thirty-six hundred years ago, Tibetans forged a new relationship with barley, when it became their central agricultural partner. Living with barley allowed Tibetans to live in a much-enlarged territory. Before this time, archaeological records show that Tibetan communities mainly lived at altitudes up to twenty-five hundred meters, which was the maximum altitude for their staple crop of millet to grow; but as the climate grew colder, they shifted from millet to barley. Compared to millet, barley helped Tibetan communities live another thousand meters higher, which dramatically expanded the places they could live. Barley allowed Tibetans to settle at altitudes up to thirty-four hundred meters, together enabling the world's highest agricultural systems.

Initially I suspected that millet must have first been domesticated in a hot-weather region and barley in a cold-weather region, even perhaps in Southwest China. I read, however, that both crops were originally domesticated in the Fertile Crescent of present-day Iran, Palestine, and Israel,[10] so it was just by chance that barley had capacities that millet, grown side by side with it, did not have. Barley seedlings were better able to handle late spring frosts, and its seeds matured more quickly than millet seeds, ripening before the early fall frosts. This significant difference is a result of biologically specific capacities in each plant; each plant makes worlds in different ways and does so in relationship to others. Barley began a focused relationship with people of the Fertile Crescent around ten thousand years ago, and somehow, along with millet, it ended up far away in the Sino-Tibetan borderlands around four thousand years ago. Along this journey, the plant transformed from its time in the Fertile Crescent as a fall-sown crop that ripened before the summer drought, to a summer-sown crop that ripened before the early fall frosts in the Himalayas.[11] Barley basically flipped its seasonality, showing its incredible flexibility.[12]

From a world-making perspective, as an annual plant, barley has gone through at least four thousand generations of evolution in this particular place, affected by climate and soil, as well as by these particular humans who are raising different varieties and selecting for certain qualities such as size, taste, ability to grind well into flour, storability, and so on. But not only is barley selected by farmers; insects and fungi also shape which varieties prosper and which disappear. Along barley's long journey, it escaped the worlds of some "pests" (a term that we often use to describe organisms who want to eat the same food we do) and entered the worlds of new ones.

One of these new ones was a kind of fungus, which Norbu, an older man with a love of cards and barley beer, told me about. His mother had told him stories of anthropologists coming to visit her, curious about Tibetan wedding rituals; he thought it funny that I didn't care much about weddings but cared a lot about mushrooms. I had brought up the issue of how Japanese officials were worried about pesticides contaminating their matsutake, so he took me to his barley fields to see how they had a disease that turns some seeds black, swollen, and powdery; he told me, "You see, we don't spray anything here. If we had pesticides, we would have sprayed this disease. It is really bad this year." I had seen it as a disease but hadn't yet realized it was a particular organism, an ergot (*Claviceps purpurea*) that was a fungus. Seen another way, what we call a disease is, like a "pest," another organism who eats what we want to eat or build houses with. Later, Norbu showed me another fungal disease that also fed on barley, this one being a kind of "damping off" that hurts seedlings (*Fusarium* sp.). We looked at the soil together, which was lighter and thinner than the soils I was more familiar with in Kunming. I learned that these soils are thinner in part because plants have less time to grow and enrich the soil with organic matter, which is mainly generated by plants' fungus-enhanced growth and fungal-digested death. As well, agriculture here is also mediated by the mountains made by the Indian tectonic plate, which not only made it colder here but also produced glaciers that more recently scrapped these lands of what soil had been created, glaciers that have since retreated back into the mountains but are still visible.

In the villages where I went to study the matsutake economy, barley is the staple food, and throughout the Tibetan region it constitutes two-thirds of the total diet.[13] People toasted their seeds for additional flavor before grinding them into a powder called *tsampa*, which is mixed with yak milk or butter. Tsampa is rolled into little balls in beautifully carved and painted wooden bowls with lids. Sometimes the seeds are fermented into a beer called *chhaang*, which provides a surprisingly strong feeling of warmth and contentment, despite its low alcohol content. After the stalks are gathered, dried, and threshed of their seeds, the stems become hay. One of the striking and unique features of many villages is the mas-

FIGURE 6.4. Large racks for storing hay over winter for yaks.
Photo by SingerGM/Shutterstock.

sive wooden racks where the hay is kept all winter (fig. 6.4). Thus, barley does double duty, nourishing humans as well as yaks.

Worlds Made with Yaks

Around the same time as they devoted themselves to the life of barley, Tibetans entered into a partnership with yaks, slowly taming the wild ones who lived in the surrounding mountains.[14] To gain a perspective on this long period of time in relationship to Tibetan history, yaks and barley were already part of the culture for about three thousand years before Buddhism appeared one thousand years ago. To grasp this in more multispecies world-making terms, I like to think of this period of four thousand years as two hundred generations of people and one thousand generations of yaks (for each female yak often gives birth at age four), and four thousand generations of barley.[15]

　　Some scholars think of yaks and Tibetans as one of the most intense entanglements between one particular animal and one particular group of people in the world.[16] Anthropologists will likely remember another famous case of human-bovine entanglement: E. E. Evan-Pritchard's 1940 account of the Nuer in Sudan and their cattle. "They depend on the herds for their very existence. . . . Cattle are the thread that runs through Nuer institutions, language, rites of passage, politics, economy, and allegiances."[17] For Nuer and Tibetans, such intimacies go far beyond mere practical survival; these bovines are not just fully instrumentalized or objectified animals for humans. Although some Tibetan nomads also raise goats, sheep, and horses, the yak stands above these other animals. Tibetan lives have been centrally organized around the qualities and capacities of yaks: as pack animals and providers of critical materials such as hair, milk, meat, and dung, they are also vulnerable beings who may need some help in calving, who attract packs of predatory wolves, and who can become ill and need medication. In the last century, some nomads can now purchase alternatives to these critical materials, but for many people here, yaks remain irreplaceable (fig. 6.5).

The world-making qualities of yaks—their ability to travel at high elevations in intense cold and thick snow—allowed Tibetans to create the tea-horse caravan. Unlike sheep or goats, yaks can carry heavy loads. Without yaks, there might not have previously been much tea in India, or it would have been incredibly expensive had human porters carried it over high snowy passes on their own backs. Yaks thrive in places that humans call "low-oxygen environments"; yaks fear low elevations and become increasingly liable to sickness when descending below three thousand meters, or around ten thousand feet.[18]

Yaks' bodies provide critical materials such as hair that can be felted for clothes or a tent or twisted into rope. Around Gyalthang, no family lives year-round in a tent, but many own one, to set up at various festivals, such as horse-riding contests when their grasslands kin travel south to compete, tell stories, reaffirm old relationships, and make new ones.

Yak dung is essential as fuel and for use in construction—such as for building walls that protect herds of yak from wolves and shelter calves during cold, windy nights. Above the tree line, yak dung is the only fuel available for keeping warm. Yaks can eat two-inch-high plants (impossible for humans to turn into fuel), creating dung that is easily dried and

FIGURE 6.5. Yaks, essential pack animals. Note the long hair on the legs.
Photo by Michail Vorobyev / Shutterstock.

burns hot. Although people with wooden homes often use an iron stove, nomads tend to build a stove at every camp with a mixture of clay and yak dung, which is then fueled with dung. Herders pay attention to how yak dung and yak milk is shaped by yak-plant entanglements. Many lactating cows are milked, and the milk is turned into butter or cheese. Herders learn what kinds of plants help yaks thrive, so that they can provide plentiful and fatty milk for calves and people of all ages (such as a kind of sedge, *Kobresia*). Herders keep close watch for what kinds of plants make the milk taste funny (like wild onions, *Allium*), what plants create thin yak bodies with little nutrition, or even which ones can thin the herd with poison. As I mentioned earlier, the Shangri-la yaks mainly eat barley hay in winter and wild plants in summer. Yaks demand that people travel with them up, down, and across mountains, alternating between summer and winter pastures.

In this high-altitude region, especially in places beyond the range of barley, yaks are what make human survival and flourishing possible. Far in the mountains there are some herds of wild yaks, which can still interbreed with domestic yaks even after millennia of domestication.

They are also financially valuable, for there is a ready market for the butter and cheese, dried meat, and hides that they provide. Historically, monks sometimes played a role in the trade; women who took care of milking the yaks and making butter and cheese would place a wheel of cheese outside their tent, and in exchange monks would leave a wheel of fermented Pu-erh tea.

Yaks don't require a market to provide a living for people, but their goods are valuable and have long been in demand. Many yak goods are storable: cheese, hides, and yak jerky. Unlike the fresh matsutake economy, with its rapid beat, the yak economy is a slow, steady pulse, inflected by the yak's slow movements and long life, often more than twenty years. When calves are born at the same time as a human baby, they can grow up together, the yaks dying just as their human companions become adults. Yaks often live through two generations of dogs, the other most important animal that is shaped, in turn, by co-becoming with Tibetans and yak. Such relations happen over the generations, as each partner learns and adjusts over centuries. Within a generation, humans gain in understanding of yaks, as well as their individual personalities. Yaks also notice human plans, have their favorite people, and tolerate having their milk taken from their calves. Yaks harbor expectations: I was told that when they catch the scent of a wolf on the wind, they turn to their protectors, either canine or human. Although wild yaks have been shaping these ecologies for millions of years, when they began living in more intimate relations with humans, it is likely that yaks populations increased, as people hunted them less and cared for them more, as peoples' lives were yakked and yaks' lives were peopled.

The Relevance of Seasonal Timing:
Yak, Barley, and Matsutake

Once a family committed itself to harvesting matsutake, it meant that these entanglements to yaks and barley were reconfigured: the entry of a new organism changed the family's relationship with all the others.[19] Likewise, the exit of an important organism also created important changes; when some Tibetans sold off their yak herds to devote themselves to finding and selling matsutake, this meant that their yearly round changed in a drastic way. For some families, this was their first year without yaks within memory of many generations. I heard of some cases where family members cried inconsolably when their yaks were sold off and handed over to someone else. I heard of grandmothers who were giving away yaks that were the great great grandchildren of yaks they had cared for when they were young. Because such entanglements require intense and repeated caretaking, and because people and yaks often create emotional bonds, these relationships are special.

Compared to the kinds of relations between yaks and humans, understanding mutual entanglements between humans and matsutake is more difficult. We cannot easily see matsutake changing its behavior in relation to human activity over the long term, let alone in the moment. Yet, if we imagine other organisms as having agency only in terms of their reaction to human presence, this is an impoverished notion of agency that defines it in anthropocentric terms.[20] Instead, if we think of the broad range of world-making activities that the species carries out every day or at some point in the annual cycle or at a certain point in its life (like mating), then we see a lot that looks like agency even if these activities don't always reach a particular threshold based on certain notions of intentionality.[21]

Unlike a deer fleeing from the hunter, matsutake do not hide from the mushroom hunter, even if that is what the people doing the hunting might think when they can't find the mushrooms whose presence they seek. From the perspective of many mushrooms, mammals can be spore vectors, carrying mushrooms' potential progeny to new places that might not be reached by the wind. For many humans, matsutake are food, but few people imagine themselves as vectors, and if we defecate into sewer systems the chances are low that we become so. Fruits are one way that plants invite animals to eat their flesh and, most importantly, their seeds. Animals become the seeds' vector for travel and fertilizer—animal friends with benefits. Mushrooms are a fungi's way to spread their spores—using the wind or the rain, or by being eaten by animals—but we know little about the role that animal transport plays in the life of matsutake.

Matsutake remain elusive not only to mushroom hunters but also to scientists trying to domesticate them, and this elusive quality is one way in which people experience fungal agency. Although so many scientists have worked for many years, spending many millions of dollars in the process, to encourage matsutake to make relations with pine trees and fruit in the lab or the forest, it seems that matsutake have refused. Despite many people's efforts to domesticate matsutake, to make them reproduce and grow where we desire them to, the proposals for domestication we have made have all been rejected by the matsutake. This is despite human's having learned about and provided for the needs of many other edible fungi, from shiitake to portabella to oyster mushrooms, mainly decomposers who eat dead matter, rather than mycorrhizal mushrooms that form relationships to living plants.

Matsutake's mating season—what is misdescribed in botanical terms as "fruiting," as if the mushrooms contain fertile seeds, but what is really a form of sporulating—takes place at the same time that Tibetans are bringing their yaks back from summer pasture, harvesting grain, and storing hay for them for the winter, as well as gathering animal bedding and collecting firewood. As plants wind down after a vigorous summer of growth, the matsutake burst forth from the forest soil, sending waves of spores into the cooling air. They are hoping to get their potential progeny into the world before the killing frosts sink deeper through the canopy. As the top layer of ground freezes, so do the matsutake. Other mushrooms can withstand a solid freeze, like the perennial shelf fungus that grows from the sides of trees, but like almost all the soil-based large mushrooms, matsutake freezes solid and then, as the sun warms the air, it turns limp and mushy, returning to the ground. Soon, there is no trace that the mushrooms were ever there. Matsutake's timing is part of their world-making and has a certain rhythm. Hunters slowly attune themselves to this beat, and to the series of temperatures and rains that might foster the mycelia to knit together a beautiful mushroom that pops up from the earth, filled to the gills with spores ready to be sent out into the atmosphere.

Industrial Forestry and the Matsutake Economy

Yak and barley generally live in open places with relatively few trees and abundant sunlight, but matsutake need forests, and this region is renowned for them. As I described in chapter 4, the crash of tectonic plates is what formed the Himalayas, and as these previously warm, low-elevation lands ascended in altitude, the climatic conditions for pines and oaks emerged, allowing matsutake to live and proliferate in this spot. For millennia, Tibetans had used these trees to build and heat their homes and to cook food. Like yaks and barley, the trees' presence made certain forms of life possible, and here, with abundant wood, yak dung was used more for fertilizer than for fuel. Keeping warm through the winter required enormous quantities of firewood.[22] Occasionally, large-scale supplies of lumber were required to build massive monasteries (such as Ganden Sumtseling) and buildings in Gyalthang, but in general, timber didn't travel far. Only in the few places where large trees grew alongside rivers, and thus were easily rafted and floated downstream, was it possible to transport lumber a significant distance.

When the scouts for the state's forestry department came to survey this part of the world, starting in the early 1960s, they did not look down at the mushrooms. Instead, they looked up and saw magnificent stands of trees, some of them more than one thousand years old. Growing slowly, with tight growth rings that lent the wood great strength and beauty, these trees motivated foresters to plan an extensive series of roads in this challenging landscape of steep slopes, monsoon-triggered landslides, and high passes that remained snow covered from early fall until midsummer.

Despite these challenging conditions, as Daniel Winkler remarks, in just forty years (1960–2000), the Chinese state removed far more old-growth forests here than the Tibetans did over the course of millennia.[23] Starting in the 1970s, Beijing set up logging camps and built sawmills, punching roads into these mountains, often at great expense. The first boom in this logging didn't last even a decade: from virtually no logging in 1976, old-growth logs started being cut and sawed, with levels of timber peaking in the early 1980s and then starting to decline.[24] Nonetheless, even after the most easily accessible trees were felled, state-organized logging continued until 1999.

During this time, Tibetans watched the state extract their territory's vast timber wealth. Some people rallied to protect groves of old trees under the care of monasteries; others joined logging crews, drove logging trucks, or carried out their own operations, often under the cover of darkness.[25] From the matsutake's point of view, however, such clear-cutting of old growth was not necessarily bad news. When slopes were not too steep and soil stayed in place, retaining enough water and nutrients to conjure up another generation of trees from their midst, and the right combination of pines and spruces emerged from the landscape, matsutake could flourish again. Logging does not always destroy mushroom habitat because matsutake can grow in the wake of massive forest transformation, especially as young pines replace old-growth forest. Even when human management fails in terms of sustainable lumber production and creates places that foresters call sites of ruin, this can lead to new habitats for matsutake's proliferation—their spores ready to land, make connections, and grow into mycelial networks.[26]

Yet in 1999, after nearly four decades of industrial logging, this economy came to a grinding halt, and many scrambled to find alternative ways to make a living. This time, foresters looked down and found mushrooms. What happened? In 1998, China's main river, the Yangtze, flooded and affected more than two hundred million people (nearly one in five Chinese citizens), making more than fifteen million homeless, and killing more than three thousand people. It not only destroyed homes but buried some huge areas of agricultural land with various kinds of sediment that are poor for growing crops. In response, Beijing started to pivot in the direction of becoming a "nature state,"[27] where landscapes were valued as much for their "environmental services" as they were for their natural resources. More specifically, Beijing started thinking about the uplands, including a lot of northern Yunnan, less in terms as a site for extracting trees and more as a watershed to absorb rainfall and reduce floods; and forests were a way to retain soil. Such a move was undoubtedly connected to waning confidence that floods could be controlled through dams. In 1998, China was almost done building the world's largest on the Yangtze—the Three Gorges Dam—but even it failed to stop the floods. Beijing's more ecological approach—working not by provincial boundaries, as it had in the past, but by watershed boundaries—was a fresh way of looking at things, but it also caused sweeping changes for residents of this upper watershed, including for Tibetans in Northwest Yunnan.

We tend to see the effects of the logging ban through a human lens, focusing on one set of beings, the trees, which are enlisted in an effort to reduce flooding. From Beijing's perspective, these trees shifted from being a way to build socialism, to being a way to make money, to becoming allies of dams and downstream communities in keeping water and soil in place. Trees became treasured for their world-making relationship to soil, their ability to cushion the fall of raindrops by absorbing them with their canopy, and their far-reaching roots that held in place steep slopes that might otherwise shear off in landslides. The life of the trees also affected other organisms such as yaks and fungi.[28]

The flood prompted sweeping policy changes, including the world's largest logging ban (covering an area twice the size of California) and massive tree-planting efforts. The floods and Beijing's political response directly suppressed other previously dominant economic livelihoods such as timber, the grazing of animals, and farming. Northwest Yunnan had almost no industry;[29] logging had been generating up to 80 percent of local government incomes. Some clandestine illegal logging continued,[30] but the state forestry centers were shut down, and former loggers, often Han from central China, were often provided with state assistance. But many rural Tibetans who were not directly involved with logging also felt the impact of the campaign strongly because it made herding their yaks more difficult.

The campaign reconfigured Tibetans' relations to yaks and barley. In terms of yaks, tree-planting efforts often took over some of the most fertile patches of grassland, and herders were now responsible for making sure their animals didn't harm these seedlings. As well, burning back grasslands to promote grazing was made illegal, and herders saw the rapid movement of trees onto grasslands as a form of encroachment. The campaign did not often directly impact barley cultivation, as barley was typically grown on level, not steeply sloping, fields, but barley hay became less valuable with diminished herds of yak. Thus, as the ban made relations between yaks and Tibetans more challenging, it simultaneously enhanced relations between matsutake and Tibetans as mushroom picking became one of the few viable options left to make a living.

The Mushroom Trade Comes of Age

Although the matsutake trade was already up and running before the ban, the ban clearly made it stronger. Matsutake's seasonal fruiting is a challenge for Tibetans because it occurs at the busiest time of the year, right when the yaks need tending, the barley needs harvesting, and the tourists begin to arrive. Unlike various other activities that adult male Tibetans may engage in, such as driving a logging truck or plowing a field, mushroom picking is an inclusive activity that the entire family can contribute to. Trudging up the surrounding hills, fathers, mothers, children, and grandparents slowly begin to carry out the silent hunt. Mushroom hunting is one of the few activities that small children can engage in to help a family make money, and they are often skilled hunters. Kids thought of it as fun, but as the season went on, day after day, some found it less so.

Some young men and women remarked that unlike working for hotels and restaurants, making money by picking mushrooms was a way to maintain far more autonomy and independence. This was mainly because they didn't have to work under a boss and also because they could do this work with friends. Young adults would go off together to hunt, often riding their matsutake-financed motorcycles, decorated with small carpets and flags. Sometimes one got to spend time with a boyfriend or girlfriend away from the watchful gaze of elders. When a couple goes out to pick, if they come back with a good haul of mushrooms, often sharing some of the money with each of their families, then there is a bit of lenience, but when they occasionally come back nearly empty-handed, as if they hadn't been focused on their task, then rumors begin to fly. Thus, matsutake hunting does seem to allow space for more forms of private courtship than in the past, but it too has its limits!

In contrast to tourism, matsutake picking was a much more equitable activity. First of all, even though matsutake grew unevenly, there were far more places to pick them in this region than there were profitable places to develop tourism. The widespread growth of matsutake was owed mainly to its own world-making powers, its ability to spread and make relationships with trees. As an activity, matsutake picking has a very low "barrier to entry." Gathering mushrooms demanded almost no outlay of money, which is important in a region where average incomes hovered around two hundred dollars a year per adult.[31]

Although matsutake hunting required little in the way of specialized gear, it did require electricity and roads. Some of the smaller villages didn't have reliable electricity, so they couldn't depend on refrigerators, which were a key part of a dealer's equipment; and a few of the more active dealers bought generators to keep their mushrooms cool. Although some dealers used horses or donkeys to travel to higher-elevation villages, all eventually needed access to roads that cars could use. Some appreciated how the tourist trade encouraged the state to upgrade the few routes that created the main tourist circuit, nicely paved roads that dealers found convenient to use.

As I've mentioned earlier, the fresh matsutake trade demanded a different kind of tempo, and thus, in some ways, a different kind of infrastructure from that required by the trade in dried goods, and this region is a critical site for that trade too. For dried goods, including medicinal plants and other mushrooms, even the most isolated villages could participate. Access to electricity was irrelevant for those who gathered these plants, which they would dry either in the sun or in a wood-powered drier. Unlike matsutake, with its insect guests and rapidly deteriorating condition, dried plants and fungi could be prepared at a calm pace, and families could dry a whole season's harvest over several months and sell it in giant plastic sacks, which were relatively lightweight and could be lashed on the back of someone hiking down a path from a village to a serviceable road. Traders in fresh matsutake (fig. 6.6), however, needed to be closer to the roads—not necessarily asphalt

FIGURE 6.6. Awaiting the opening of the daily matsutake market.
Photo by Michael J. Hathaway.

roads, but at least cobblestoned ones that did not turn to thick mud during the rainy season—a problem that could go on for months.

In one village I lived in, there was such a road. The high level of craftsmanship on this road, and the way it enabled people to come and go even during the monsoons, earned this road much appreciation. Previously, one had to walk home several miles from the main paved road, sometimes carrying heavy loads of supplies in the pouring rain. As the rainy season progressed, one would walk in an increasingly rutted muddy path along the road, until a new one was made, but there was sometimes little room. Villagers had been petitioning the government to pave their side road for years, but officials said they needed to concentrate funds on outlying villages that did not yet have any vehicle-passable roads, and that they would have to wait. Finally, the village leaders rallied everyone to pool funds and make the road; later on, villagers referred to it, partially in jest, as "the road built by Japanese stomachs."

On the other hand, many admit that the mushroom business also had a number of downsides. The matsutake market takes place during a short window of time, which overlaps with the end of the main tourist season. The market doesn't always provide as much income as sellers would like, especially as the price has gone down compared to when their parents were selling in the late 1980s. One day while waiting for the market to open, as the several laoban drove up to the market site and villagers squatted or stood in small groups chatting, I got into a conversation with a young man and his grandfather. The old man was remarking what a time it was, about twenty years ago, when his grandson was born, and when people felt rich with matsutake money. It was a tremendous discovery that the Japanese loved this one mushroom that Tibetans picked occasionally but didn't especially like to eat. He said back then, the matsutake was like gold, laughing at his new term "mushroom gold." Except in this case, he said, matsutake were like magic, the veins appeared, year after year renewing themselves, unlike the veins of gold in the ground that, once dug out, were gone forever and left mining pits scattered across the land. He heard that to the south, people were digging many holes in the forest, looking for truffles buried underground. "Now *truffles* are more like gold. When the truffles are gone, the truffles are gone, and all we have are a bunch of holes."

The Social Impacts of Matsutake Money

Resurgence with Houses: The Rise of the Matsutake Mansions

In Northwest Yunnan, I was intrigued to see that the matsutake money was stimulating an ethnic resurgence that was reminiscent of the changes it brought about for the Yi, but with some important differences. Whereas for the Yi, the most obvious manifestation was the rise of Yi-centered cultural spaces for the entire community, among Tibetans, the most prominent result was the rise of spaces for one family, what mycologist David Arora deemed "matsutake mansions."[32] These

FIGURE 6.7. The exterior of a matsutake mansion.
Photo by the Escape of Malee / Shutterstock.

massive structures had intricate Tibetan Buddhist carvings on the outside (fig. 6.7) and elaborate paintings on the inside, especially lining the
walls of the main guest room where everyone ate and drank tea (fig. 6.8).
These houses were most often found in villages that happened, by luck,
to be situated within a morning's walk of rich matsutake habitat, and
they seemed to pop up almost as quickly as the mushrooms
themselves.

These mansions are truly monumental structures and are even more
impressive from the inside. I had assumed that many Tibetan artists
were being employed by this now widespread use of matsutake money
but soon learned otherwise. Sitting down to tea around an old iron
wood-burning stove, an elderly man informed me that the painters are
mainly "rats." He could see the confusion on my face and then, with a

FIGURE 6.8. The interior of a matsutake mansion. Photo by Pete Oxford / Danita Delimont / Alamy.

bit of a grin, said they were Sichuan-ren: people from Sichuan. I later realized this meant Han Chinese. This was quite surprising to me: Sichuanese were being trained and were spending vast quantities of time creating elaborate Tibetan religious paintings. Tibetans, in turn, have a kind of pleasure in hiring Han to do religiously specific work for them—that is, to do their bidding. Such a frisson is generated especially because of the long legacy of Han (often conflated with the Chinese state) as serting their control over Tibetan lives, and particularly in their restrictions on Tibetans' religious expression. When Tibetan home owners hire ethnic Bai people—talented wood carvers from the town of Jianchuan, where more than twenty thousand people are engaged in the wood-carving business—to create windows and doors for them, it doesn't have nearly the same kind of emotional charge as hiring Han. Tibetans view the Bai as another ethnic minority that is relatively high up within ethnic hierarchies in Yunnan, but the Bai are not seen as a source of their historic oppression.

The "hiring of rats" speaks to one of the more powerful social effects of matsutake wealth: within their local area, the majority Han, who mainly live in the larger towns in prime matsutake territory (and occupy almost all the governmental and professional positions as well as making up the vast majority of store owners), now have to reckon with Tibetans as serious economic players. Ethnic Han often feel a sense of cultural superiority, and there is often a conflation of being Chinese with being Han. With the exception of two dynasties during China's long period of imperial rule, almost all the major rulers have identified as Han,[33] and since the end of imperial rule in 1911, the central government has been almost completely dominated by Han leaders.

Han and other non-Tibetan townspeople have dismissed Tibetans as dirty, backward, and poor; but now Tibetans have new economic clout, and this change is powerfully experienced by local Tibetans. Thus, their pride not only is taking place within their villages, in the creation of these powerful symbols of their wealth and culture, but it also has altered their dealings with Han in town. The growing possibilities present in Yunnan have also attracted Tibetan capital from the transnational diaspora of Tibetans in India and Nepal and beyond—in places as far flung as the United States and Switzerland. Wealthy Tibetan traders, who earned profits by exporting Tibetan rugs and other goods abroad, are bringing back wealth and investing it in hotels and tour businesses.

Sometimes, however, matsutake money wasn't invested in things or people in ways that led to long-term improvement, and I learned that this wealth also led to problems. With the help of Norbu's translation, I talked to some of the older women in villages around Gyalthang who didn't speak Mandarin. They showed more concern than I had heard from the men. They said that their new homes and vehicles were the most important purchases made. Some worried that their husbands were squandering the money by gambling, which has increased with these new flows of cash, and they were glad to have invested in some major purchases like a house or vehicle that would not just be easily gambled away. They told me how matsutake money is different from yak money, that even though the mushrooms are picked by the whole family, some men insist that they should control these profits. Men have been the ones who would usually get the earnings from selling yaks and other big-ticket items. They are the ones who typically head out, cash in hand, to buy a truck. They are usually the ones who travel far distances to study in monasteries. Other kinds of money, they say, like the kinds made from selling yak butter or cheese, are more often kept by the women, the ones who are milking the yaks, especially when the milk comes on strong in the summer months when the meadows are buzzing with life. Sometimes the women say to their husbands that matsutake money is more like milking a cow than selling it; it is more properly money than should go to them, not to the men. As one woman said, "Sometimes the men listen, and sometimes they don't. We don't gamble with the money; we try to save it." They often encouraged the family to put money into building a new house, which represents a relatively safe and permanent investment. Yet as a few women pointed out, even though they were proud of their homes, houses cannot easily create wealth, unlike investment in vehicles. And so, with a sense of cultural pride and fearful reluctance, some women encourage their husbands or sons to gather up all the money, after years and years of picking mushrooms, and join the modern caravan, or use the wealth in supporting their family's religious connections.

Resurgence of Religious Practice: The Rise of Monastic Training and Pilgrimage

Especially since the beginning of the PRC in 1949, Tibetans have experienced strong opposition to their Buddhist religious practices, which historically shaped their economy and everyday life in nearly every conceivable way. Around three decades later, however, as the matsutake economy gained ground, people talked about a major shift in China's political winds, indicating Beijing's greater religious tolerance. In 1983, the state funded the rebuilding of Ganden Sumtseling Monastery and gave permission for the Panchen Lama—the second-highest lama after the Dalai Lama—to visit. His visit was still clear in the memories of several elders I talked with, and it seemed to signal a significant shift in allowing important religious figures greater freedom of movement. This led to many conflicted feelings, with some Tibetans being surprised, others grateful, and some suspicious. As mentioned earlier, those with suspicions remembered just thirty years earlier when Tibetans were promised room to practice their religion, but as their lives were eventually restricted, there were two armed rebellions that resulted in the almost one-sided slaughter of tens of thousands.[34] Despite decades of a kind of settler colonialism that promoted secularism and punished forms of religious practice, many Tibetans remained committed to their faith and noted that, despite an occasional climate of greater religious tolerance, Beijing still carried out periodic crackdowns.[35]

Matsutake wealth also enabled families to fund their own activities, such as sending children away for religious study and undertaking long-distance pilgrimages; these acts generated stronger connections between Tibetans, sacred landscapes, and temples across vast stretches of land that had long been part of religious circuits. This led, as well, to a greater connection with Lhasa, a place that had been off most local people's itineraries for some time, and which remained an arduous bus trip that could take five days. The funds and the expanded sense of regional affiliation also reinvigorated linkages with monasteries in places like Qinghai, where long-existing connections had remained a bit stronger. I met elders who had studied in Qinghai as young men, and they described how their grandsons were starting to go there again, whereas their sons had not been able to. Matsutake wealth helped enable them to carry out these long-distance connections, fostering a sense of greater kinship with their fellow "tsampa eaters," a term sometimes used to describe connections among Tibetans who might be affiliated with different Buddhist teachings or speak different dialects or languages. The last transformation that I will discuss is yet another matsutake-enabled one that allowed Tibetans to reclaim their historic role as long-distance traders—this time not with yaks but with trucks.

Resurgence of the Ancient Caravan, This Time with Trucks

One thing I hadn't anticipated was the way that matsutake funds facilitated investment not only in massive houses but also in trucks—which are regarded as a kind of reassertion of Tibetans' historic role as traders in the tea-horse caravan. Unlike driving a tourist minivan, which some Tibetan men felt might compromise their dignity, long-distance truck driving was an activity that many of them saw as particularly masculine. This was a livelihood requiring bravery, skill, and endurance: As there are few rest areas and repair shops along the roads, drivers need to be able to fix their trucks when the inevitable breakdowns occur. The long switchbacks are dangerous, and brakes can easily overheat. Plus frequent landslides mean that sizeable rocks can come down the slopes at any moment.

Driving along some of these roads without safety fences, where one can see the river far below and occasionally spot rusting heaps of trucks, was a constant reminder of the dangers inherent in this work. Truck drivers might go on long journeys, such as delivering goods to Lhasa, which, if traveling by bus, is hopefully a four-day trip with alternating drivers. Like bus drivers, some truck drivers traveled in pairs so they could take turns, or if traveling alone, the driver would pull off the road to rest. This was all in a social context of male bravado where many would try to stay awake as long as they could. Sometimes I would overhear a group of men in their thirties sitting at a table bragging about the vast distances they could cover between sleeps, and how they negotiated periodic landslides, as well as recalling stories of their friends who perished.

There is a certain pride, then, among some Tibetan long distance truck drivers not only that they have the bravery to negotiate these dangerous mountain roads, and have success in doing so, but also that they are reviving their heritage as traders. A number of Tibetan truckers who got their start from matsutake money proudly told me they were bringing the caravan route "back to life." I also heard similar sentiments from older men who saw the vigorous matsutake trade as being related to their Tibetan past. They didn't see themselves as people at the edge of the Chinese empire but as traders along ancient transnational routes. These truckers sometimes drove alongside Han matsutake dealers with their SUVs literally stuffed to the gills with mushrooms, and as we will see, these dealers can be helpful.

How Matsutake Dealers Are Changing the Roads

Matsutake mushrooms appear only during the rainy season, which creates major challenges for Northwest Yunnan's road infrastructure, as I mentioned above, with its many roads cut into steep mountainsides and therefore causing many landslides. These challenges are a frequent topic of matsutake bosses. Often late at night, sitting in a smoky room while the matsutake laoban lords it over his loyal staff, men slightly younger than him, almost all with a cheap cigarette, play cards. Sitting there in a Tibetan village reminded me a bit of the Yi gatherings hosted by their matsutake dealers, except up here, higher in the mountains, there were more stories of disasters on the road. All told, in several different locations, I heard a dozen tales of dealers rushing to the airport with a truckload of matsutake, only to be blocked by a massive landslide (fig. 6.9). About half the time, there was nothing to be done: *mei banfa*, literally "no method" or "no way." If this was an isolated road and the rains kept coming, sometimes they waited there for three days, walking for hours to get cell reception. In these tales, dealers would sometimes delight in their own tragedy, as their cargo—what had been a fortune—slowly turned to worm food. Coal trucks just wait it out, a few days mean nothing other than hunger and boredom; but for mushroom dealers, a long wait means disaster.

In other stories, the drama mounts when the laoban is stopped by landslide and he begins to take action. Walking around the bend to find a signal, he calls his *guanxi* (social connections) network. In one story, the matsutake laoban knows someone who knows someone who has the cell phone number of the local road-crew boss, who is awakened from a hangover at 3:00 a.m. The road-crew boss brings a handful of his coworkers, who, with an old bulldozer and explosives, improvise a way through the massive slide of mud and boulders. In a few hours, all is passable for a few thousand yuan. Sometimes the matsutake van or

FIGURE 6.9. One of the frequent landslides in the rainy mushroom season.
Photo by Shuang Li / Shutterstock.

truck can't even drive over the humped, rocky, or mud-slicked path. In this case, a thick, rusty cable is wrapped under a low-geared, boulder-dented *Dongfeng* (east wind) truck, the workhorses of Chinese develop-ment, which arrives on the other end of the slide.

The dealers, less fearful if they have their own skilled driver, can be white-knuckled, trying not to peer down almost a thousand feet into the valley beneath the road with no railings, and forget to breathe as their wheels occasionally spin and their vehicle lurches forward. They are half pulled over the massive mound of muddy rock but also need to steer to keep their vehicle on the newly made path. These are heroic tales, and while some end in tragedy and some in victory, both reinforce the sense that these matsutake laoban are fearless entrepreneurs and risk takers, deserving of their wealth.

These are not tales of individual physical effort, but rather tales that stress the power of the bosses' social connections: It's not just the carrot of a monetary reward that brings the road-crew boss out of his warm bed to drive all night to a disaster along a steep road. The matsutake boss knows that the road-crew boss gets out of bed because he is close to the caller; their lives are entangled in years of banter, shared drinks, mutual assistance, jokes, and more. The road-crew boss knows that this is also a dangerous place, for sometimes a road crew that is clearing one slide becomes victim to another slide that buries them under a new pile of debris or pushes them over the side to the valley below. Yet unlike most of the work they do, which is done because they have to, these moments off the clock at someone's personal request can be especially rewarding; they accrue a special kind of merit as a favor done for a friend and not just as a job that must be done.

Of course, people do all kinds of things for money; many people will risk their lives on a regular basis. My point is not that such risks are by any means unique to the matsutake economy, just that the mushroom's own world-making matters, and that it fruits in relationship to a weather pattern and within a very particular infrastructure. As well, the fragility of the matsutake's body after being picked makes a big difference. Coal can be stockpiled for years, but fresh matsutake decays quickly. Here, close to the equator, matsutake can grow only in high-altitude locations, which means that the terrain is steep and the roads are dangerous. In other parts of the world, farther from the equator in places like Sweden, matsutake grow in a relatively flat landscape; and driving on paved roads in Sweden is simple and safe. Thus, this particular configuration in Yunnan includes how the mushrooms' lives intersect with the particulars of the terrain, climate, infrastructure, and economy that has been built around them, as humans, barley, yaks and many others make a living together in this place.

Conclusion

As we have seen, matsutake's specific requirements and traits, its unique world-making qualities, have engaged the lives of many rural Yunnanese Tibetans, and in so doing, their entangled worlds have changed. Just forty years ago, Tibetan economies revolved mostly around the lives of barley and yaks—as they had done for nearly four thousand years. These two organisms were critical to Tibetan livelihoods and remain important, since few other things can survive the challenges of these high altitudes, and thus replace them.

But I have not wanted to present these places as timeless, nor the relationships as necessarily harmonious and symbiotic (between certain people, a certain plant, or a certain animal); instead, I've tried to show how a long history of intense strife has shaped this place and these beings that have lived here. Some say that the introduction of Buddhism here—with its prohibition against killing animals, which is widely but not always followed—meant that fewer yaks were killed and the Tibetan diet shifted away from meat and more toward milk. These historical changes inflect the relations between organisms, and likely, with fewer yak killed for meat, the herds grew. The shift from millet to barley, which took place nearly thirty-six hundred years ago, allowed Tibetans to greatly expand their range of settlements into higher elevations in this vast upland terrain.

The beginning of the matsutake economy in the 1980s—what some older Tibetans called the "mushroom gold" that never runs out— offered a strikingly new way to make money in a place where options remained few. The excitement of these initial decades, when prices increased and fortunes were made, is easily remembered by older Tibetans who were adults when this happened. Some are waiting for an upturn in this economy to happen again, while others say that the time when Japanese stomachs developed their region and built their roads is finished. Yet in 2017, matsutake once again became a precious item in Yunnan when, for various reasons that are not well understood, matsutake emerged in only a few places. As Japanese wholesalers can now buy from over a dozen countries around the world, conditions in Yunnan do not set the global market price; on the other hand, as Yunnan's importance has grown, so has its influence in Japan. In 2017, the prices shot up to three thousand yuan (about US$443) per kilogram,[36] whereas in 2006 the prices were about twenty yuan (about US$2.50) per kilogram. Keep in mind that in 2017, China's poverty line was about thirty-five hundred yuan (about US$518) per year,[37] so finding an armful of matsutake that year could lift one out of poverty for an entire year. Some places were blessed with productive patches; others had nothing. That year matsutake brought great wealth to a handful of pickers and very little money to many.

Since 1999, the new environmental turn taken by the PRC to foster the growth of trees in the Yangtze River's upper watershed has had the unintended effect of making the land less productive for yaks. In places where yaks have left, the wild grassland plants quickly began to change—revealing one particular aspect of yak world-making—how their eating habits affected the landscape. Over the past four thousand years, these landscapes have been made more by the efforts of yaks than by those of humans; yaks' cumulative actions have changed the ecology, mainly through the plants they have eaten and the plants they have avoided.[38] Of course, yaks did not make this place themselves; they did so together with human care, with human-lit fires that increased grasslands, human-raised and human-dried barley hay for winter food supplements, and with human protection from predators and more. Weathering a recent series of booms and busts, from Japan's economic bubble to the rise and fall of China's industrial forestry, yak and barley remain foundational to the Tibetan economy.

But for others, the push of the now twenty-year-long policies for stopping logging, encouraging tree planting, and criminalizing grassland burning now combines with the pull of the mushroom. Some worry about the threats posed by this newfound matsutake wealth, as yuan are lost in gambling and alcohol or in projects that don't pan out, where money is squandered, or worse, turns family members or neighbors against each other. Others are reminded that matsutake have enabled them to build houses like never before, and to forge stronger connections throughout Tibetan territory and reclaim their place, once again, as masters of the caravan that connected them to the wider world more than a thousand years ago. Like before, they do not do so alone, but with intentional allies and unintentional participants in their lives, lived with and in tandem with yaks, barley, dogs, matsutake, and a number of others.

One morning, catching the scent of burning boughs of juniper offered to Gandha, the goddess of smell, and seeing that nearly everyone
was preparing to hunt mushrooms, my Tibetan guide, Norbu, turned
to me and told me that the day before, he had been talking to one of
his new friends about yak, and matsutake came up. His friend said that he
planned to keep his yak, as they were so dear to him, but he knew his
neighbors were getting ready to sell. "I sometimes think the yak are
getting jealous of the matsutake, we give so much care to them." This
idea, that the yak are jealous of matsutake, speaks to the kinds of entanglements that all organisms have, sometimes with and sometimes
without humans.

CHAPTER SEVEN

Final Thoughts on Understanding Fungi and Others as World-Makers

What if we shifted from a world where nonhumans are regarded as resources or machine-like automatons to a world where humans and nonhumans alike are considered "persons," fellow beings who are actively carrying out their lives? What does it mean to see the yeast that we wake from its slumber to enlist its powers in making bread, the spider that lives and hunts in the corner of the room, the crows that search for food in the middle of our cities, all as fellow beings who are actively carrying out their lives?

I notice the matsutake popping up from a carpet of rich moss, and as I pry it from the ground, I wonder what the matsutake makes of me. I realize that I am just one of the many organisms with which this mushroom makes worlds. More than a thousand years ago, some person in Japan, likely entranced by the smell of the matsutake, bit into it, not knowing if it would be nourishing or deadly. Now we know.

—Reflections on fieldwork

Over the past decade, my life has become rich with fungal entangle-ments, and I find myself being remade through these relations. For years, my interest in fungi was mainly oriented toward a handful of wild edible mushrooms: morels, chanterelles, and chicken-of-the-woods, and I enjoyed walking in the woods to find them. But my increasingly varied and complex encounters with a myriad of fungi made me ever more mindful of the numerous forms that they can take (from mushrooms to lichens to molds and yeast) and sparked my curiosity about diverse and powerful ways of fungal being.

At times, my curiosity borders on the obsessive. I should have had a premonition of this during my first research trip to Japan with Anna Tsing and Shiho Satsuka, two fellow members of the Matsutake Worlds Research Group. Shiho Satsuka made an appointment for us to meet with Dr. Ogawa. We were excited, as in Japan he had been a student of Dr. Hamada, one of the world's first scientists to conduct laboratory and field-based research on matsutake, and together they had been publishing their work since the 1960s. In the United States and Canada, we had a hard time tracking down articles from these early days of Japanese matsutake research, as almost none of these articles had been digitized, and few of them were published in journals available in North American university libraries. Dr. Ogawa had very recently retired, and we imagined that he would be just the person to help us find a treasure trove of documents since he would surely have a critical matsutake archive.

When we met, he shared our enthusiasm for our project of understanding the world-traveling career of matsutake, but when we asked him about the archives, his face dropped, and he looked down as he said, "Ahhhh . . . if you had only been here last month—I just threw four file cabinets' worth of papers into the trash. What a pity!" We let out a collective gasp.

Only years later did I gain an empathetic view of what happened. I now think that maybe the reason why he let go of his life's work was to free himself, in some small way, from the grasp of the fungi, who had, some might say, advanced their own lives through his fungi-promoting scholarship and activities. I, too, sometimes feel that the fungi are both literally enabling me to digest the world that I eat and at the same time metaphorically turning me into their substrate, their food, even as I advocate on their behalf.

Although I have been actively studying mycology for the past seven years, I still count myself a beginner in fungal worlds, as I increasingly realize how much there is to know and to learn. When I first began to walk through the woods to purposefully seek out fungi, I remember being mainly interested in the few mushrooms I could confidently identify as delicious and not poisonous. Later, I began learning about mushrooms' other powers and their diverse roles in the forest, and I found this even more fascinating. On mushroom walks, I encourage others to notice the smellscape of the forest, which makes up the aromatic equivalent of an orchestra. Some can learn to pick out the different notes in the fungal symphony. We roll logs over to see the networks of mycelia at the boundary between wood and soil, glimpsing the wood-wide web under our feet. We get right up close, engaging in the kind of close looking that botanist Robin Wall Kimmerer describes when talking about mosses, as being almost like listening: attuning one's whole body to discern fine details and distinctions.[1]

On one such walk, someone catches a ray of light at just the right angle to illuminate clouds of spores billowing out from artist's conk fungus (*Ganoderma applanatum*) on a large Douglas fir tree (*Pseudotsuga menziesii*). The shelf fungi are not just there; they are actively exploring the tree, slowly converting its heartwood into soft crumbly blocks. As they do so, they are extracting nutrition from the wood, breaking down the cellulose and lignin like no other kingdom of life can do. The tree, in turn, notices the fungi and tries to build a wall of strong wood cells to isolate the fungi from spreading.[2] Eventually, some insects will be attracted to the smell of this fungi-infused and nutrient-enriched wood, and they will attract woodpeckers that will transfer wood-decaying fungi on the feathers near their beak.[3]

In turn, another fungus arrives, this time in special baskets called mycangia that are built into the jaws of the mountain pine beetle (*Dendroctonus ponderosae*). When the beetles lay their eggs in the living tree, they also seed the fungi from their mycangia, which grows and transforms the once hard wood, too tough for larval mouths, into a material that is soft and nourishing. There are so many of these kinds of intimate relations between fungi and different animal species—interspecies relations that are not always transmitted in the DNA but are nonetheless handed down from adult to offspring, like this beetle with its basketful of mushrooms that it passes to its young. Other interkingdom relations are not handed down from generation to generation but are made anew. For example, as we saw, each matsutake spore must seek out its most critical life partners, such as a particular kind of tree, and make relationships that fuse them together, perhaps joining in an extensive network that stretches throughout the forest.

What I find interesting to contemplate about these kinds of interspecies relationships is that the whole set of underlying assumptions and norms in the discipline of biology would be different, as the mycologist Alan Rayner has suggested, had it taken a symbiotic organism like a mycorrhizal matsutake mushroom and not a bounded individual, like a rat, as a model organism to generate the basic theories of life.[4] But biology was not modeled on mycorrhizal relations between fungi and trees as the basis for how the world works, but rather on studies of organisms presumed to be individuals, whether the rats put in sterile cages to understand human psychology or fruit flies raised in vials to understand genetics.

In this book, I have invited you to think about certain questions, such as: What might it mean to eschew the kind of anthropocentrism that has dominated Western modes of thought for centuries? How might we find other ways of approaching the world? Many scientists now see other animals as using tools, planning for the future, and communicating—abilities that were previously assumed to be solely possessed by human beings. I welcome this growing recognition of other animals' abilities even as I challenge the expectation among many that these abilities cease at the animal border. What happens when we go beyond what could be called animal centrism?

What would it mean to recognize a greater diversity of bodies and ways of living and of active world-making that extend into the realm of plants, fungi, bacteria, and others? For one, we might move beyond our animal-centered expectations about what counts as movement, communication, perception, and agency, and we could do this, in part, by not defining these things so narrowly that only humans and other animals can meet the criteria. At the same time, I have wanted to show that the notion of world-making can help us avoid the opposite extreme by defining agency in ways that might be too general and too broad. Following Jane Bennett or Bruno Latour, one can see all things (living and nonliving) as expressions of "vibrant matter"[5] or as "actants,"[6] playing seemingly equal roles. From Bennett's or Latour's viewpoint, however, there is no interest in the different abilities and agencies of a tick, an elephant, or a bacteria. Thus, while both "actant" and "vibrant matter" approaches challenge assumptions of human exceptionalism, they do not foreground curiosity, exploration, and knowledge about the lives of different beings. This kind of grounded interest in the specifics of other organisms' lifeways, perceptions, and engagements with others is part of what I am calling a "lively approach."

The notion of world-making is thus one kind of a lively approach, which draws on the work of many scientists, such as Uexküll, to understand the unique and specific worlds made by a wide range of organisms. Many other scientists have also challenged the notion that nonhumans are passive and mechanistic, and they are part a "minor tradition" within the otherwise Cartesian Western sciences. Indeed, even Charles Darwin was part of this minor tradition, before his theories were retooled by the neo-Darwinists. Scholars such as Carla Hustak and Natasha Myers, Gillian Beer, Jessica Riskin, and Elizabeth Grosz found that Darwin himself cultivated a deep curiosity about the liveliness of animals and plants.[7]

This book opened with a vision of how fungi made our planet green, and I will end with a sense of how fungi will continue to do so. I am confident that some kinds of fungi will continue to thrive even if the world, through humans' own actions, becomes uninhabitable to not only humans but many other species as well. I have often heard two particular stories that aim to explain the rise of a "damaged planet."[8] The first is that our successful domination of the planet is just an inevitable by-product of the categorically more advanced intelligence that we have over all other animals, which led to the rise of technology, from the stone hammer to the nuclear reactor. As you have likely detected from this book, I am suspicious of such claims, in part because they are based on an assumption of human exceptionalist intelligence, and in part because I do not celebrate such dominance as an achievement.

The second story, which I see much more merit in, advances the notion that the root of our ecological problems lies in a version of global capitalism that spurs extraction, production, and consumption without limits. This economic mode not only disconnects consumers from the conditions of production of the goods they use, and workers from the products of their labor; it disconnects each of us from the places where the plants, animals, minerals, and fungi are harvested, killed, gathered, or mined. My initial trip to China in 1995 was motivated by my desire to live in a place that boasted at least three decades of vigorous socialism (1949–79). But already, long before I traveled there, China was well known for its environmental damage,[9] despite its leaders' earlier claims that pollution only occurs under capitalism.[10] China's notable failing in this regard led me to conclude that the North American capitalism I was raised in and the Chinese socialism I learned about shared three powerful orientations: human exceptionalism, human supremacy, and resourcism. These orientations are linked to the Enlightenment project and its aftermath. They have had a particularly powerful structuring effect on political and social institutions, no matter their economic basis.

Resourcism as a mindset became much clearer to me at a workshop organized by my colleague Ling Zhang. She gathered a group of scholars and asked: How did living and nonliving things become "resources?" So many of us had been trained to see the world as a compilation of "natural resources" whose potential value for human projects seemed always already present. The philosopher Heidegger gave insight into resourcism with his remark that it entails "viewing the natural world as a standing reserve" for human utility.[11] This mindset is so deep that many environmentalists use the language of resourcism when talking about "conserving" or "protecting" a "natural resource." Such language ignores other beings' own life projects and world-making.

There are alternative perspectives, some that are early in the process of becoming, and some that have strong ties to the deep past. Many people throughout the world were never completely convinced by the project of Enlightenment, with its rational disenchantment of the non-human world and how it turned others into resources. One alternative is the "rights of nature" movement, which has often been Indigenous led or at least informed. It first started as a national force in Ecuador and has spread to New Zealand, Bolivia, Sweden, and Austria; indeed, its growing acceptance is evidenced in the fact that by 2012, some 147 countries had recognized some form of the rights of nature.[12] In part, this occurred through the creation of legal structures that recognize rivers and other places as "persons."[13]

Throughout the world, many creative and energetic people are working to change legal and economic systems that have been built on resourcism. Unraveling these centuries-old systems and building alternatives will take a lot of rethinking and a lot of work and will be met with deep resistance by the institutions that these systems benefit from.

It is my hope that this book might be one small piece of this larger movement, and I join with many other thinkers who are showing us how we live in worlds created by other lively beings, whose actions far exceed the expectations of mechanistic theory.[14] Together, we are proposing new ways to understand biology, challenging exceptionalist and supremacist orientations, and hoping to find other ways to understand and act in the world as we move into the future. My journeys into the fungal world give me a strong sense that fungi's diverse lives and their world-making efforts offer us insights into this kind of future. As a form of life often engaged in intense symbiotic relations with others, fungi—existing as a lichen on a rock or tree, a mycorrhizal fungus living underground, or endophytic fungi living within plant leaves—help us challenge the premise that the world is full of individuals, rather than made and remade through relationships. As Donna Haraway puts it, "What happens when human exceptionalism and bounded individualism, those old saws of Western philosophy and political economics, become unthinkable in the best sciences, whether natural or social?"[15]

Years ago, when I became a member of the Matsutake Worlds Research Group, I envisioned my goal as understanding how the matsutake commodity chain was built by people. This led me back to Yunnan Province for several research trips starting in 2010. These experiences transformed the very quest itself, as it widened to encompass understanding the role fungi played in making our world, as well as fungi's own efforts at world-making. Discovering these fungal activities has led me to think about how the often invisible and inaudible fungal actions or movements or performances all add up. Fungal actions are a form of cumulative agency that helped enable plants to leave the oceans and join fungi in exploring forms of terrestrial life. Fungi helped push plants up into the atmosphere, where they could mine carbon from the air, sequestering some of it into their bodies and sharing some with their critical life partners. Aboveground, fungi are also eating plants, living and dead, and belowground they are eating rocks, as nutrients are once again cycled through living beings until they return again to the earth's great oceanic sinks. Since I began this project, the estimated number of fungus species on the planet has grown, so that the already small percentage of identified fungi has proportionally decreased over time. Many of the new ones discovered are not being found deep in the heart of rainforests or on remote mountain peaks, but in places right under our very noses. As I have learned more about fungi, I have increasingly noticed that I no longer need to go into the woods to see them. I have started to recognize that fungi exist in a spectacular range of forms, that they are everywhere, living on all kinds of surfaces, inhabiting all kinds of substrates.

My own encounter with a kind of interesting urban fungus happened quite by chance. For years I had noticed many small stains on the concrete sidewalks around my house but I hadn't paid them much attention. Then one day while looking down, I got on my knees to see what these stains were, and another world opened up for me. What had looked before almost like a spot where chewing gum had been removed turned out to be a lichen, the same organism I introduced in chapter 1, which has been living on the planet since long before the days of the dinosaurs and which helped turn the rocky surface into soil, giving plants a habitat to use. Whereas matsutake live as a symbiotic organism with their mycelia and a network of trees, and their fruiting bodies persist for only a few weeks a year, lichen are symbiotic organisms (fungi and an alga or a photosynthetic cyanobacteria) that have likely been living on my sidewalk for decades, walked on by many thousands of people. Nonetheless, they are still going strong and could live on the concrete long after the sidewalk has crumbled and the pieces are left in a heap. After first noticing lichens on the sidewalk, I found them elsewhere in the middle of the city: clinging to the walls of buildings, digging their rhizines into wooden fences, coating rusty pieces of metal in beautiful patterns, and so often living on tree trunks that many people assume they are simply the bark. I recently noticed that they can grow on top of each other, and I found the lichen equivalent of a lucky clover, a lichen layer cake stacked four levels tall. As I looked at more street trees, I noticed different kinds of lichens from those that eat the sidewalk; they make different relations to different kinds of trees. Some trees (such as maples) support an abundance of mosses and lichens, while others (such as cedars) seem to repel these life forms. I was taught to identify maples by their leaves, but now I identify them by their trunk, especially when it is mossy and lichen-rich. As maple trees grow older, more mosses, lichens, and sometimes ferns arrive and flourish on this new island of habitat—without the tree, the lichens would not have been there, so trees foster the presence of lichen and moss worlds.

I have begun to notice a lot of fungal life inside buildings, as well, and I feel differently about them than I used to; for example, when I find molds growing on my shower curtain, my knowledge has changed my relationship to them. Previously, I was only repulsed by them, but now I am surprised to find that some are beautiful colors—for example, the color of Spanish wine. I also began marveling at their apparent ability to live without food; it seems logical that mold can prosper on old oranges and bread, but surprisingly, with the addition of only water and bits of food (like soap, skin flakes, or grout), mold can thrive on glazed ceramic tile. For fungi, our homes are potential habitat and substrate. The water we bring into our home through showers or boiling food or our own respiration and perspiration can bring forth fungi and encourage their omnipresent spores to germinate. Even in places where thick timbers have been pressure-treated with poisonous arsenic designed to kill all forms of life, fungi have turned them into powder. At my own university, although builders used concrete and not wood, in part to avoid the potential for rot to destroy the structural timbers, there are sections where the spread of mold is now widespread, necessitating massive renovations and causing millions of dollars of damage. Some mold even found a home in sound-absorbing panels in my school's gym. Although such panels do not normally become a fungal substrate, someone noticed that the panels absorbed moisture from the air, from the evaporation of sweat from me and thousands of other athletes. Fungi don't like to eat dry food; they require a drink, and this moisture provided all they needed. These are reminders of fungi's power to grow and thrive in unexpected and unwanted places. Engineers and architects would do better to consider fungi as active world-makers, as their designs can create substrate for fungi to thrive in unintended ways.

If you try to find examples of how different organisms play a part in world-making, you might find, in some biology textbooks, the term "animal engineer" or "animal architect"; and the most likely case study you will encounter is the beaver. Historically, fungi have been far less visible than beavers as shapers of landscapes. Fungi have rarely been accorded the status of architect or engineer, but this is starting to change. Indeed, my first chapter would not have been possible without a whole slew of fantastic new stories by scientists regarding their explorations of our fungal planet. I was delighted that some of these scientists were my fellow Vancouverites, including Suzanne Simard, whose work I described in chapter 3, and some of her students have now attended my talks. Conversations with Margaret McFall-Ngai helped me understand that fungi are not the sole victims of biologists' discrimination; indeed, all microbes have been tarnished by the brush of the germ, which began a "pathogenic bias" that still dominates the field of microbiology—a realm that is defined more by the human eyes (and our incapacity to see these tiny organisms) than by a shared biological relationship.

I was also motivated and enlightened by neighborhood friends such as Willoughby Arevalo and Mendel Skulski, who have read and thought deeply about the world's fungal presence. Further south, in Washington, lives the world's most famous fungal evangelist, Paul Stamets, who is a major force behind a mushroom renaissance. In one lecture, one of the world's most watched TED talks, he promotes the idea that "mushrooms can save the world." Stamets advocates mushrooms as medicine for humans and honey bees, and as a means to clean up pollution. One time he noticed oyster mushrooms growing from a pile of oil-contaminated soil, which led him to experiment with ways fungi could offer a form of "mycoremediation." He and others are motivating a younger generation of mushroom enthusiasts to carry out new creative experiments in DIY mushroom cultivation in their homes and to clean up polluted soil and water.[16] One experimenter trained an oyster mushroom to consume a diet made up almost entirely of cigarette butts, an extreme form of mycoremediation that hints at fungi's broad powers to break down toxic substances.[17] Together, these thinkers might help usher more people toward a different view of the web of life—a view that shifts our own ways of world-making by discovering and acknowledging fungal presence and power.

I am excited by what I am hearing about this "mushroom turn." Even in England, home to the culture that promoted "toadstool kicking," exhibits and lectures on mushrooms are selling out. In particular, I am interested in this expanding vision of working with fungi, and I am intrigued by the notion that "mushrooms can save the world." But I admit, I do have some caveats, especially if this is approached with an attitude of "fifty ways to exploit mushrooms." And I am especially leery of a "silver bullet" approach that assumes we can carry on business as usual when it comes to poisoning the earth (intentionally or otherwise) with the assurance that fungi will clean up the mess. While they might be able to do something, fungi will unlikely be a singular panacea, and such claims also perpetuate a resourcist attitude that I wish to challenge.

More recently, a plastic-eating fungus, *Aspergillus tubingensis*, was discovered in a landfill in Islamabad.[18] I had assumed that plastics could not break down into anything other than micro-particles, and thus the buildup of plastics was inevitable. I had not understood fungi's wondrous alchemical abilities to turn xenobiotic (i.e., human-created and chemically novel) substances into food. What was humbling, though, was to think about the possible scale of time. "Sure, fungi will eventually eat plastic," said Willoughby, "but that might take tens of million years or so, just like what happened with lignin so long ago."

European scientists (and before them "natural historians") had previously regarded fungi as a strange form of plant, bound to the soil, an organism that moved slowly over time, if it moved at all. For centuries, fungi have been mainly seen as things that lived solely on the ground or on trees. Yet as we turn to observe the air, we find many fungal spores, often in bewildering assortments. We realize that with every breath, from our first to our last, we bring new fungi into our bodies. Some spores are fragile, while others are incredibly tough. Many spores land just a few inches from the mushroom cap; others are caught on the wind and float over the highest mountain ranges, sometimes entering the jet stream and crossing continents. With respect to water, we have only recently discovered dizzying levels of spores in seawater, freshwater, and tap water.[19]

Spores appear and germinate in places where we least expect them, such as inside spaceships ostensibly sterilized as human-only spaces. Fungi have been found in places thought inhospitable to all forms of life, for example, in Chernobyl's still-hot radioactive core. There, they not only survive but have discovered how to turn radiation into food.

These radiation and plastic-eating fungi provide evidence of fungi's continual and creative world-making efforts. Fungi are forming new relations with nonorganic materials, creating new pathways and flows of energy that perhaps had not previously existed on the planet. Although fungal digestion is often depicted as a passive act of absorption, these fungi not only create a special recipe of enzymatic chemicals; they also use their mycelial strength to rip apart the molecules of plastic, physically tearing the plastic into pieces, the better to digest it.[20] This is not merely a case of "discovering" a new form of food, but actively devising the means to convert something into food. Such creative work in decomposing the world, a form of world-making, is a fungal specialty. Fungi live their lives in ways both strange and familiar, as they connect the flows of the living and the dead, as they cross kingdoms of life, and as they foster diverse networks of connection. They live within us, on us, and all around us, making worlds with many other organisms, and we can start to glimpse and understand some of these worlds.

Some fifty years ago, one of the world's most prominent anthropologists, Claude Lévi-Strauss, declared that "animals are good to think with";[21] by this he meant that references to animals are a rich source of symbolic language, providing insights into different cultures. I have tried in this book to say the same and more about fungi, in that, by deeply considering some fungal worlds, we can gain further insights into different kingdoms of living beings, and thus into life on Earth writ large. This book is born out of a deep love of and respect for scientific knowledge, which has taught me much about this kingdom. Yet on the other hand, the course of this research has revealed some of the cultural and historical underpinnings, and limitations, of Western science. I have also shown how fungi are much more than just good for humans to think with, for they have, in fact, made thinking itself possible for humans and all the other terrestrial forms of life that emerged after fungi's early explorations onto land helped make the planet green. Fungi are necessary for us to live; they are critical life partners not only for mycorrhizal plants, but for all living beings, including us. This recognition allows us to see fungi not just as a commodity for exploitation but as a hidden source of life and a continuing wonder that challenges some of our animal-centric, anthropocentric, human exceptionalist, and resourcist ways of understanding.

My time in China and Japan, places that are unencumbered by a history of mycophobia, showed me other ways to see fungal alliances. In China,

I met many people who loved all kinds of mushrooms and who were keenly aware of their presence, and many were deeply involved with medicinal mushrooms. Their enthusiasm helped China become one of the world's most important centers for research on medicinal fungi—a level of prominence built on a legacy dating back nearly two thousand years. Today, China is reclaiming its ancient status as generating the greatest number of fungi-inclusive medicinal texts.[22] It has also recently become a major site for research on matsutake's medicinal qualities in response to a rising global fascination with, and increased consumer demand for, medicinal fungi.[23] As Kunming-based researcher Zhou Dequn told me: "For years, dealers in China suspected the Japanese—why did they pay such a high price for matsutake when they didn't taste very good? We thought the Japanese were secretly using them for medicine. So we started our own research, to explore matsutake's potential."

In response to this avalanche of interest, I advocate a world-making perspective to challenge the resourcist and anthropocentric tendencies that have dominated discussions of our use of other organisms. A world-making approach asks: How do each of these mushroom species experience their own lives, and how are they in relation to others? Take, for example, chaga (*Inonotus obliquus*), a recent fungal medicinal obsession: whereas studies of matsutake mushrooms have not shown conclusive evidence that picking reduces their numbers, in part because they live underground where they are often hidden and also because of how they grow, the life of chaga is different. As a slow-growing perennial, it is visible to hunters every day of the year, and therefore much more likely to be completely eliminated from a forest. There is growing concern that chaga might become rare or endangered, and while this is important, I would also like us to extend our concern to chaga's wider ecological engagements. Chaga eats birch trees of a certain age, so removing mass quantities of this fungus from a forest might impact not only chaga but also birch, as well as the many other forest organisms that interact with them. Many reports proclaim chaga's benefits to human bodies, but almost no research explores how chaga influences the forest, through engagement with insects, animals, birds, and trees. In short, there is no exploration of chaga's own world-making.

In Japan, I met scientists who offered me a vision of the natural world that did not advocate human exceptionalism or a competition-based model of biology. Anna Tsing and Shiho Satsuka showed how American forestry scientists largely adopted a "wilderness protection" model of forestry, which assumes that human actions can only ever damage ecosystems.[24] In contrast, Japanese forestry scientists see humans as part of the environment and human actions as having the potential to be environmentally neutral, damaging, or beneficial. As Tsing describes in *The Mushroom at the End of the World*, "'raking' symbolizes disturbance in both sciences—but with opposite valences." In the United States, raking is seen as destroying matsutake, by disturbing underground fungal bodies.[25] In Japan, raking is seen as making productive matsutake forests, because it uncovers mineral soil for pine trees to grow and reduces duff favored by other fungi who can out-compete matsutake. Thus, North American foresters experimented with rules to limit pickers' inevitable damage, whereas Japanese foresters experimented with how raking could foster matsutake's growth.

In Japan, I saw widespread public interest in fungi. A pamphlet produced by the Mycological Society of Japan introduced me to the concept of fungi as "life partners." The pamphlet used the term to refer specifically to human-fungi relations, and it talked about the ways that laypeople can observe living fungi, such as shitake mushrooms, bread yeasts, and koji (used to make soy sauce, miso, and sake). I have expanded this term to include fungi's nonhuman allies—what I have been calling their "critical life partners," such as various trees and insects. The society also created a guide—see the appendix—to one hundred fungi that were voted most "closely linked to the lives of Japanese people." When will we create a list of the fungi most closely linked to people's lives in Europe or North America?

Another example of fungi and microbes being part of pop culture is Masayuki Ishikawa's wonderful Japanese manga (graphic novel), translated into English as *Moyasimon: Tales of Agriculture* and introduced to me by my MWRG colleague Shiho Satsuka.[26] In this story, a young boy named Tadayasu has incredible powers—he can see, understand, and communicate with microbes, as he watches them carrying out their lives. He attends agriculture school and troubleshoots problems when people try to work with microbes to brew sake or to make soy sauce and miso paste. Reading this was the first time that I saw microbes not being portrayed in cartoonish form as "menacing germs." In one scene, a matsutake mycelium is depicted with its hyphae talking to each other, chagrined but also proud, saying something to the effect of: "We are busy working underground to build the Sagrada Familia (an ornate cathedral designed by Antoni Gaudi), but humans don't even know we are here! Architects indeed."[27]

Japan is not only the center of the world's matsutake market; it is also the center for scientific knowledge about matsutake cultivation. Thousands of studies have been published in Japanese on its growth, and relatively few of these have been translated into other languages. Japan hosts dozens of research stations working on matsutake, some using field experiments and others working in high-tech labs. I have seen many researchers fall under matsutake's spell, as this seemingly simple fungus evades the efforts of some of the smartest scientists using some of the most advanced technologies. Many billions of yen have been spent, often with the goal of eventually cultivating the mushroom, but so far it has eluded these efforts. Scientists have successfully inoculated pine tree roots with matsutake mycelia but have not been able to get these mycelia to transform themselves into a mushroom. While such reports continue to pile up, the conditions under which matsutake form fruiting bodies still remain mysterious. As Burkhard Bilger's *New Yorker* essay on matsutake says: "They've sequenced most of [matsutake's] genome, replanted cuttings from host trees, inoculated seedlings with cultures, and experimented with fertilizers. Yet the mycelia never survive transplant. 'It's a much more difficult challenge than truffles. . . . There is just something that the tree provides that we don't,' Lefevre said."[28]

Some people see matsutake's unwillingness to fruit as a form of "mushroom resistance" and evidence of fungal agency, but this is a limited way of understanding agency. As I have shown, matsutake do not merely exhibit agency when they refuse human desires; they show it through everyday world-making. Matsutake continues to surprise scientists, spurring them on to new projects. In China as well as Canada and elsewhere, even when matsutake prices are low, these mushrooms continue to pull people out of bed on frosty mornings, to put on cold, stiff boots and go on a quiet hunt.

Shiho Satsuka is writing the third and final book in our matsutake trilogy; it will take you into the world of Japanese farmers, scientists, and others who work with matsutake so you can learn through their experiments and attunements. Shiho, Anna, and I have traveled throughout Asia and the Pacific Northwest, listening to farmers, mushroom hunters, dealers, and scientists, and smelling and watching the mushrooms themselves and the animals and plants they attract. And, in the process, we have become mushroom people.

I will end with a brief story about the travels of a lichen. After my immersion in this project, I started to notice pieces of rock covered in lichen, some with bright-yellow splotches and multicolored puzzle-like formations. I happened upon an especially beautiful orange-lichen-covered rock near my friend's home in the mountains of New Mexico. When I picked it up, it was still warm from lying on a slab of sandstone. I brought it home to Vancouver and kept it outside, thinking the lichen would prefer the direct sunshine of the outdoors and not the muted light within my home; even so, four years later I noticed it was fading. Slowly, slowly, it was no longer what it used to be. I wrapped the rock carefully in a box and mailed it back to my friends. They placed it in their yard and within a year reported that the strong colors had returned—the lichen had regained vibrancy after breathing in the same air that it had evolved in over deep time, perhaps even using melanin to convert some of the sun's radioactivity into its own food. This lichenated rock is doing many things at scales of space beyond what I am able to see, and at scales of time that I cannot observe, and yet the work of many scientists, with their attunements and their machines and their deep curiosity, has opened up this world for me. I know that this rock is also breathing out oxygen through its algal companion, oxygen that I have used myself, and that its rhizines are slowly eating away its own home, so slowly that a million years later, together they may weigh virtually the same, but the lichen will have converted the rock into biological life. Maybe then, some human will pick it back up again, appreciate its beauty and its unique world-making powers. Then again, likely no one will, but no matter: with or without human presence, this lichen and its kin will, as they have been doing for hundreds of millions of years, continue to turn rocks into food, keep this planet green, and bring an abundance of life into being.

APPENDIX

TABLE A.1. One Hundred Selections of Fungi Voted Most Connected to People in Japan

Number	Japanese Name	Scientific Name	Notes
1	Egg bamboo	*Amanita hemibapha* (Berk. & Broome) Sacc.s	A representative mushroom that colors summer in Japan.
2	Ibotenus mushroom	*Amanita ibotengutake* T. Oda, C. Tanaka & Ts	Although it is a poisonous mushroom, it has a strong flavor component called ibotenic acid.
3	Fly agaric	*Amanita muscaria* (L.) Lam.	Representative of fairy tale without mushroom.
4	Amanita mushroom	*Amanita pantherina* (DC.) Krombh.	Also known as fly agaric.
5	Egg amanita	*Amanita phalloides* (Vaill.ex Fr.) Link	A killer mushroom wearing a green umbrella.
6	Doctrine mushroom	*Amanita viros* a (Fr.) Bertill.	Its neat appearance is a deadly poisonous mushroom worthy of being called an "angel of death."
7	Water shimeji	*Ampulloclitocybe clavipes* (Pers.) Redhead, L	Poisonous mushrooms that are incompatible with sake.
8	Yachihirohitake	*Armillaria ectypa* (Fr.) Emel	A globally endangered species produced in Japan.
9	Naratake	*Armillaria mellea* (Vahl.) P. Kumm.	It is a wild mushroom that represents Japan and has various faces.
10	Aspergillus niger	*Aspergillus niger* Tiegh.	Used for industrial production of organic acids and enzymes.
11	Kikouji	*Aspergillus oryzae* (Ahlb.) E. Cohn	Indispensable for Japanese fermented foods, the basis of Japanese taste.
12	Hyuga hunting chicken Hyuga bamboo fungus	*Asterinella hiugensis* Hino & Hidaka	The extinct Japanese endemic mold.

(*continued*)

TABLE A.1. (*continued*)

Number	Japanese Name	Scientific Name	Notes
13	Horsetail	*Astraeus hygrometricus* (Pers.) Morgan	A star that has landed on the ground.
14	Asunaro tengu's nest fungus	*Blastospora betulae* S. Kaneko & Hirats.f.	The first fungi named by the Japanese. Original name from Caeoma asnaro Shirai (1889).
15	Kurokawa	*Boletopsis leucomelaena* (Pers.) Fayod	Foodies have a famous autumn taste.
16	Yamatake mushroom	*Boletus edulis* Bull.	In Europe, the king of wild edible mushrooms along with Trif.
17	Kirinomitake	*Chorioactis geaster* (Peck) Kupfer	An endangered species whose production area is extremely limited.
18	Camellia kinkakuchawantake	*Ciborinia camelliae* LM Kohn	Occurs on petals of camellia and sasanqua that have become sclerotium, signaling the coming of spring.
19	Cladosporium and cladosporioi	*Cladosporium cladosporioides* (Fresen.) GA	The most common airborne germs in Japan.
20	Doksasaco	*Clitocybe acromelalga* Ichimura	If you eat, you will suffer from hell, and you will be terrified.
21	Choleforma empetri	*Coleophoma empetri* (Rostr.) Petr.	Produces antifungal precursors from Japan.
22	Hita toyotake	*Coprinopsis atramentaria* (Bull.) Fr.	The umbrella melts overnight, and its name manifests itself.
23	Kingfisher	*Cordyceps heteropoda* Kobayasi	The familiar cordyceps, which is generated from the larva of the cicada.
24	Chrysanthemum	*Cordyceps militaris* (L.) Link	Cordyceps occurring from moth pupae.
25	Amanita mushroom	*Cryptoporus volvatus* (Peck) Shear	What is the origin of the name, Hitokuchi?
26	Usukikinugasatake	*Dictyophora indusiata* f. *Lutea* (Liou & L. Hwan)	An endangered species that is beautiful in both shape and color.
27	Koyakumannenaritake	*Echinodontium japonicum* Imazeki	A Japanese endemic species (endangered species) collected in Nikko, Tochigi Prefecture.
28	Sorairotake	*Entoloma aeruginosum* Hiroe	A full-body blue gem.

Number	Japanese Name	Scientific Name	Notes
29	Kusaura amanita	*Entoloma rhodopolium* (Fr.) P. Kumm.	Main member of annual mushroom poisoning.
30	Urabeni hotei shimeji	*Entoloma sarcopum* Nagas. & Hongo	A representative player of autumn thickets.
31	Licorice mushroom	*Fistulina hepatica* (Schaeff.) With.	The shape is just the liver, what is the taste?
32	Enokitake	*Flammulina velutipes* (Curtis) Singer	[Wild?] mushrooms are quite different from cultivated ones even under the snow.
33	Turtle mushroom	*Fomes fomentarius* (L.) JJ Kickx	Mushroom used as a crater.
34	Cholera mushroom	*Galerina fasciculata* Hongo	The deadly poisonous mushrooms and poisoning symptoms are similar to cholera disease.
35	Red fox	*Galiella japonica* (Yasuda) Y. Otani	Endangered species endemic to Japan.
36	Mannentake	*Ganoderma lucidum* (Curtis) P. Karst.	It is useful as a lucky charm.
37	Rice stupid disease fungus	*Gibberella fujikuroi* (Sawada) Wollenw.	Bacteria producing the plant hormone gibberellin.
38	Maitake	*Grifola frondosa* (Dicks.) Gray	An edible fungus that makes you want to dance if you find it.
39	Yamabushitake	*Hericium erinaceum* (Bull.) Pers.	Rabbit-shaped mushrooms have a new sense of texture and flavor.
40	Sakura shimeji	*Hygrophorus russula* (Fr.) Kauffman	Rather than cherry color? Edible wild mushrooms called by various names.
41	Bitter mushroom	*Hypholoma fasciculare* (Huds.) P. Kumm.	Be careful even if you grow a lot.
42	Kuritake	*Hypholoma sublateritium* (Schaeff.) Quél.	A common edible fungus similar to nagartake.
43	Bamboo shoot	*Hypomyces hyalinus* (Schwein.) Tul. & C. Tul.	This strange thing is a mushroom that has undergone mold infestation.
44	Bunashimeji	*Hypsizygus marmoreus* (Peck) Bigelow	Mushrooms that are familiar in supermarkets where cultivation methods have recently been established in Japan.

(*continued*)

TABLE A.1. (*continued*)

Number	Japanese Name	Scientific Name	Notes
45	Otter mushroom	*Inonotus mikadoi* (Lloyd) Gilb. & Ryvarden	A new species from Japan dedicated to the Emperor.
46	Spider mushroom	*Isaria atypicola* Yasuda	Koishikawa Botanical Garden is listed as a type locality in the Meiji era.
47	Tsukutsukuboushitake	*Paecilomyces cicadae* (Miq.) Samson	Cordyceps, a hint of immunosuppressants.
48	Ivy mushroom	*Lactarius lividatus* Berk. & MA Curtis	Representative of excellent wild edible fungi distributed in Japan along with pine.
49	Red fir mushroom	*Lactarius laeticolor* (S. Imai) Imazeki ex Hong	Popular with wild edible fungi along with mushrooms.
50	Chichitake	*Lactarius volemus* (Fr.) Fr.	Mushrooms especially liked in Tochigi Prefecture.
51	Onifsube	*Lanopila nipponica* (Kawam.) Kobayasi	A volleyball ball, a chimpanzee, or a demon bump?
52	Everyco	*Laricifomes officinalis* (Vill.) Kotl. & Pouzar	The Japanese name comes from the Ainu language. Ainu people used this mushroom for medicinal purposes.
53	Red fox	*Leccinum extremiorientale* (Lar. N. Vassiljeva)	A big and irregularly cracked umbrella is a landmark.
54	Shiitake mushroom	Lentinula edodes (Berk.) Pegler	Worldwide edible fungus from Japan.
55	Dust mushroom	*Lycoperdon perlatum* Pers.	If smashed, spores will fly out like dust.
56	Hatake shimeji	*Lyophyllum decastes* (Fr.) Singer	Shimeji which grows in the village.
57	Shaka shimeji	*Lyophyllum fumosum* (Pers.) PD Orton	Is Buddha surprised too? Occurred all together.
58	Hon shimeji	*Lyophyllum shimeji* (Kawam.) Hongo	A true shimeji of "Katsumatsutake Shimeji." Finally, artificial cultivation succeeded.
59	Squidfish	*Macrocybe gigantea* (Massee) Pegler & Lodge	A giant mushroom is standing in the middle.
60	Rice blast fungus	*Magnaporthe grisea* (TT Hebert) ME Barr	An important disease fungus of rice, continuous research is desired.

Number	Japanese Name	Scientific Name	Notes
61	Amigasatake	*Morchella esculenta* (L.) Pers.var. esculenta	Ascomycetes telling the coming of spring.
62	Yam mushroom	*Mycena chlorophos* (Berk. & MA Curtis) Sac	Luminous mushrooms that became a tourist attraction.
63	Usukibano mitake	*Mycena luteopallens* Peck	A delicate mushroom that grows on beech seeds (seed) in autumn.
64	Olpidium viciae	*Olpidium viciae* Kusano	Japanese were involved in elucidating life history.
65	Tsukiyotake	*Omphalotus japonicus* (Kawam.) Kirchm. & O.	Beech forest representative, mushroom poisoning No. 1, also known for luminescence.
66	Honey mushroom	*Onygena corvina* Alb. & Schwein.	A rare bacterium that occurs in the owl's beak.
67	Penicillium citrinum	*Penicillium citrinum* Thom	Produces world-class precursors for cholesterol-lowering drugs.
68	Faffia rodzima	*Phaffia rhodozyma* MW Mill., Yoney. & Soned	Industrial production of astaxanthin, a natural pigment from Japan.
69	Nameko	*Pholiota microspora* (Berk.) Sacc.	Edible mushrooms whose cultivation method has been established in Japan.
70	Fycomikes nitens	*Phycomyces nitens* (C. Agardh) Kunze	Fecal fungi that respond to light.
71	Suginotamabaritake	*Physalacria cryptomeriae* Berthier & Rogerso	A very small mushroom that grows on the dead leaves of a cedar branch.
72	Sugihiratake	*Pleurocybella porrigens* (Pers.) Singer	Sudden transformation from edible bacteria to poisonous mushrooms?
73	Tamagitake	*Pleurotus cornucopiae* (Paulet) Rolland var.ci	Natural products are now increasing in valuables and cultivated products.
74	Oyster mushroom	*Pleurotus ostreatus* (Jacq.) P. Kumm.	A typical edible mushroom in Japan, familiar at supermarkets.
75	Kaentake	*Podostroma cornu-damae* (Pat.) Boedijn	Do you suffer from flames if you eat?
76	Tamachoreitake	*Polyporus tuberaster* (Jacq.) Fr.	A rare mushroom that occurs on the sclerotium.

(continued)

TABLE A.1. (*continued*)

Number	Japanese Name	Scientific Name	Notes
77	Koubudo	*Pseudotulostoma japonica* (Kawam.) Asai, H.	A rare bacterium whose classification at the phylum level has recently been corrected.
78	Lycopodium	*Psilocybe argentipes* K. Yokoy.	Hallucinatory mushrooms (magic mushrooms) specially made in Japan.
79	Houki mushroom	*Ramaria botrytis* (Pers.) Ricken	Watch out for unique edible bacteria but similar species (poisonous bacteria).
80	Risomucor pussils	*Rhizomucor pusillus* (Lindt) Schipper	Revolution of cheese production, production of mucor rennet.
81	Shouro	*Rhizopogon rubescens* (Tul. & C. Tul.)	Premium edible mushrooms from the pine forest on the coast.
82	Rhizopus nigricans	*Rhizopus nigricans* Ehrenb.	Widely distributed in Japan, producing soybean fermented food tempe.
83	Rhodospolidium toluroides	*Rhodosporidium toruloides* Banno	The first yeast found in Japan to be a basidiomycete.
84	Shogenji	*Rozites caperatus* (Pers.) P. Karst.	The Japanese name is derived from the name of the temple.
85	Nisekuro hearts	*Russula subnigricans* Hongo	A virulent bacterium that is [Japanese?] for "I'm sorry to make a mistake."
86	Kotake mushroom	*Sarcodon aspratus* (Berk.) S. Ito	Despite its grotesque appearance, it is a delicious mushroom that goes well with Japanese food.
87	Suehirotake	*Schizophyllum commune* Fr.	Widely distributed. Contributing to the development of genetics.
88	Fox noyari	*Scleromitrula shiraiana* (Henn.) S. Imai	An endemic species of Japanese mulberry that is endangered.
89	Amitake	*Suillus bovinus* (Pers.) Roussel	It is widely distributed in pine forests and is popular as a wild edible fungus.
90	Hanaiguchi	*Suillus grevillei* (Klotzsch) Singer	Excellent edible fungi representing larch forest.

Number	Japanese Name	Scientific Name	Notes
91	Sakura tengu's nest fungus	*Taphrina wiesneri* (Ráthay) Mix	It is an important fungus that is the key to the research on the phylogenetic evolution of fungi.
92	Great white mushroom	*Termitomyces eurrhizus* (Berk.) R. Heim	Tropical mushrooms grown by termites.
93	Trichoderma harzianum	*Trichoderma harzianum* Rifai	A strong enemy of shiitake mushroom cultivation, but also for industrial production of enzymes.
94	Kawaratake	*Trametes versicolor* (L.:Fr.) Quél.	Widespread distribution of anticancer drugs due to immunostimulatory activity.
95	Bakamatsutake	*Tricholoma bakamatsutake* Hongo	"Matsutake" that accidentally appears in the grove?
96	Matsutake	*Tricholoma matsutake* (S. Ito & S. Imai) Singe	King of edible fungi, leading the world in mycorrhizal fungi research.
97	Jumping shimeji	*Tricholoma muscarium* Kawam.	Delicious Japanese specialty mushrooms, but overeating is poisonous.
98	Kishimeji	*Tricholoma ustale* (Fr.) P. Kumm.	Poisonous mushrooms ranked in the top 3 of mushroom poisoning.
99	Iboseisho	*Tuber indicum* Cooke & Massee	This is the first type of truffle found in Japan.
100	Tamanori iguchi	*Xerocomus astereicola* Imazeki	A peculiar ecology endemic to Japan, which occurs on cuticles.

Source: Mycological Society of Japan, http://www.mycologyjp.org/~msj7/WI_information_J/100.htm, visited January 18, 2021.

Note: I saw only an English version, and some of the translations are imperfect; some adjustments for formatting have been made.

NOTES

Foreword

1. *Matsutake Worlds*, eds. Lieba Faier and Michael Hathaway, Berghan, 2021.
2. Anna Tsing, *The Mushroom at the End of the World*, and Shiho Satsuka, in preparation.

Preface

1. This mushroom is called matsutake (pronounced mah-sue-tah-kay) in Japanese, sōngróng (松茸, "pine antler's velvet") in Mandarin Chinese, *be sha* (oak mushroom) in Tibetan, and *mene* in the Chuxiong dialect of Yi. In this book, I use the Japanese term as this has become the scientific species name and because Japan is the world's center of passion for this mushroom. Like the English word *deer*, *matsutake* is used as the singular and the plural.
2. As opisthokonts, we mainly inhale oxygen and exhale carbon dioxide, and we eat food, sometimes other animals, rather than synthesize it from sunlight or from chemicals.
3. Matsutake Worlds Research Group 2009a; see also Matsutake Worlds Research Group 2018.
4. Bunyard 2013.
5. Kirksey and Helmreich 2010; Ogden, Hall, and Tanita 2013; van Dooren, Kirksey, and Münster 2016.

Introduction

1. Gruen 2011.
2. Gruen 2015, 41.
3. In terms of Euro-American literary traditions, an alien might be surprised that while children's books are full of speaking nonhuman animals, they almost completely disappear in adult's books. A few notable exceptions include books such as George Orwell's *Animal Farm*, Jack London's *White Fang*, and Richard Adams's *Watership Down*. The whale in Herman Melville's *Moby Dick* doesn't speak, but he certainly has "personality." Recent books include Jamie Bastedo's *Nighthawk* and John Valliant's best-selling book *The Tiger: A True Story of Vengeance and Survival*. A recent Pulitzer Prize–winning book in the category of "botanical fiction" is *Overstory* by Robert Powers. His success challenges the advice that a mentor gave to my partner, a fiction writer, to "remember that nature cannot be a character in fiction."

Fungi have rarely been important, but when foregrounded they often play menacing roles, such as in stories by Ray Bradbury ("Boys! Raise Giant Mushrooms in Your Cellar") and John Wyndham (*The Secret People*). Others, however, have magical roles, such as in *Alice in Wonderland* (Lewis Carroll) and *Fantasia* (Joe Grant and Dick Huemer). For an overview, see the Botanical Fiction Database, https://www.thefishinprison.com/botanical-fiction-database.html. I am curious about other literary traditions around the world.

4. There are now dozens of definitions of a species. For one excellent overview of the debates, see De Queiroz (2005). I was originally going to avoid the term "species" in this book and replace it with "organism" for several reasons but ran into some challenges. The term "organism" was invented to challenge the notion of animals as machine-like, and to argue that animals' forms of living are "organized" (Keller 2008; Cheung 2010). One problem with a mere replacement of terms, however, is that "organism" doesn't distinguish between related kinds as can the term "species." For now, I reluctantly use "species," for simplicity's sake (and occasionally the term "species complex" for matsutake, as there are several closely related kinds and there are not clear distinctions between them).

5. Kemmerer 2006, 10–11.

6. Kemmerer 2006.

7. Shapiro 2008, 7.

8. The effects of the Enlightenment are global and powerful, but perhaps overstated. Thousands of non-European epistemologies and philosophies may eschew human exceptionalism (Davis and Todd 2017).

9. Hustak and Myers 2012.

10. Riskin 2016.

11. Riskin 2016.

12. Crist (1999) 2010, 5.

13. Masson and McCarthy 1996; von Frisch (1927) 1953; Cline 2010.

14. Tsing 2012; Tsing 2015; Haraway 2003; Haraway (2007) 2013; Plumwood 2002a; Myers 2015; Todd 2014; Todd 2016; De Waal 2008; Rose 2011; Lien 2015; Lien and Pálsson 2019; van Dooren 2019; van Dooren and Rose 2016; Hird 2000; Hird 2004; Despret 2016; Kimmerer 2013.

15. Jussieu and Brongniart 1851, 21.

16. Money 2011.

17. Pringle 2017.

18. Chen 2017.

19. Millman 2019, 147.

20. As Zaac Chaves pointed out to me, this split-gill fungus (*Schizophyllum commune*) is perhaps the world's most widespread fungus. From one specimen, spores are only 25 percent compatible with each other whereas spores from different specimens are 99.98 percent compatible, which reduces inbreeding. Chaves used the work of Tom Volk: https://botit.botany.wisc.edu/toms_fungi/feb2000.html.

21. King 1997.

22. Bar-On et al. 2018.

23. Talbot et al. 2008.

24. Millman 2019.

25. Ruisi et al. 2007.

26. Ivarsson, Bengtson, and Neubeck 2016.

27. Golan and Pringle 2017.

28. Ironically, Carol von Linné used Latin names in the late 1700s with the idea that each species would have a permanently affixed designation, but these names have been changing quite quickly, especially for fungi. The taxonomic history of the North American matsutake begins in 1873 with the description of *Agaricus ponderosus* by Charles Peck, which he changed five years later to *Agaricus magnivelaris*. Nine years later, Pier Saccardo transferred it to another genus and deemed it *Armillaria ponderosa*. In 1949, Rolf Singer transferred it to *Tricholoma ponderosum*, which was accepted for thirty-five years, until Scott Redhead in 1984) concluded that the correct name for the species is *Tricholoma magnivelaris* (Peck Redhead). Thirty-three years passed before Steve Trudell and others (2017) separated out *Tricholoma murrillianum* as another species.

Other fungi share a similar fate, a kind of taxonomic pinball. We might expect an occasional switch of the species names, but with affordable and quick DNA analysis, which opens up studies of kinship written in the genes, some are moved into another genus, or even placed into another family.

29. In this book I avoid a term that I have seen used hundreds of times to describe the movement of fungi and other organisms into new places: "colonize." I have seen few alternatives, and almost no one recognizes this issue in print. For me, it was only in 1993, when studying biology with a friend named Jess, that I heard someone question how the term "colonize" is rampant in biological discussions. Jess, who is Indigenous Modoc from northeastern California, pointed out that using the term colonize in biology serves to naturalize colonization as something that every species does. In 2019, when I raised the issue with mycologist Willoughby Arevalo, he had already planned to avoid the term in his book on mushroom cultivation (2019). In searching for alternatives, I noticed that the biogeographer Jonathan Sauer spoke of plants' "territory" ([1988] 1991), which included not just expanding, but retracting and morphing. Likewise, Sarah Besky and Jonathan Padwe talk about plants' territoriality, but their account emphasizes human agency in moving and using plants (2016).

Although there is fascinating research on how gendered metaphors influence scientific understandings (Martin 1991; Martin 2001; Mortimer-Sandilands and Erickson 2010; Roughgarden 2013; Bagemihl 1999; Giffney and Hird 2008), I have not yet found an analysis about how colonialism influenced biological theories. The closest I have come is the work of Banu Subramaniam, who points to language in the descriptions of "invasive plant species" that seem to repeat racist, xenophobic, and anti-immigrant discourse in the guise of scientific objectivism (2001).

30. Lorimer 2007; Satsuka 2013.

31. Bavington 2010.

32. Cheney and Seyfarth 1990.

33. Baldwin and Schultz 1983.

34. Crist (1999) 2010.

35. Mahmood 2011.

36. Probyn-Rapsey 2018.

37. Latour 1992.

38. Rather than assuming a human monopoly on power, Latour explored how scientific networks are built through connections between actants. He argues that scientists do not just produce knowledge on their own. For example, he shows how a scientist like Louis Pasteur needed to build a network with actants such as microscopes, microbes, and sheep in order to create "his" discoveries.

39. Bennett 2010. Whereas Latour and Bennett don't distinguish between the living and the nonliving, I refer here to the specific projects of living beings. Following the insights of Kim TallBear and many other scholars, I recognize that the distinction between the living and the nonliving is often blurry and may be less clear than once imagined, but I focus more on the living as widely understood (TallBear 2011).

40. Bennett and Loenhart 2011.

41. Heidegger 1995. Other scholars, such as anthropologist Hugh Raffles (2020) and idiosyncratic intellectual Roger Caillois ([1970] 1985), explore the worlds of stones, even though they are typically seen as fixed or inert. In a more strictly scientific vein, Paul Gillen's beautiful essay introduces us to Robert Hazen's concept of mineral evolution, the idea that the earth is diversifying in geological ways, with new kinds of minerals developed over time (Gillen 2016). This places minerals in a livelier and more active story of change than I've seen before, where many minerals have a specific history. Previously, I imagined that all minerals had existed since the beginning of planet Earth's formation.

42. Berlant and Warner 1998, 558; Nakayama and Morris 2015.

43. Tsing 2018. Thus, world-making is a much more encompassing concept than "point of view" or "perspective," for perception is just one aspect of world-making. The idea of a point of view maintains the notion of an outside singular world that all beings inhabit, just with different perspectives.

44. Cadena and Blaser 2018; Escobar 2018.

45. One of the few exceptions is a book by anthropologist Eduard Kohn, *How Forests Think*, which explores the notion that other beings have their own ontologies (2013).

46. In a number of Indigenous understandings, other animals are considered "persons," with their own personalities and complexities.

47. Wolf (1982) 2010.

48. Tsing 2015, 292.

49. Faier and Rofel 2014, 364.

50. Haraway (2007) 2013, 244.

51. Hehemann et al. 2010.

52. This possibility was opened up by Rosling and colleagues' 2003 study in Sweden, which was the first detailed study of ectomycorrhizal fungi throughout the soil column. Previous studies focused only on the top organic layer (Finlay 2008). They found that, counter to the previous belief that fungi would live only in this top layer, nearly half the fungi lived in this "less hospitable" mineral layer. Studies like this continue to find fungi growing in places previously unexpected.

53. In this book I will use a number of terms like "breathing" that are often used only for humans or for animals, whereas plants are often described as "respiring" and fungi are rarely described as doing either. I also describe fungi as performing, as creating actions, and as having

practices, whereas normally they are said to just "grow," a term that seems automatic. I do this deliberately to provoke reflection on the ways our language conventions foster an implicit and inescapable notion of human exceptionalism.

Chapter One. Fungal Planet: The Little-Known Story of How Fungi Helped Foster Terrestrial Life

1. S. E. Smith and Read 2008. Indeed, everything about plants—how they evolved and continue to live—has been shaped by a vast fungal presence, and vice versa, through encounters over deep time.

2. As with all scientific claims, consensus is rare, and this is especially true for events that happened long ago. There are several theories about why the age of oil-making ended. In choosing among these contending versions, I tried to select ones that seem to have rich bodies of evidence and much scientific support, but I must also reveal my bias toward looking for interesting possibilities of fungal powers.

3. For example, the term "evolutionary arms race" was invented during World War I by Julian Huxley to describe how predator and prey seem to coevolve ever more intricate capacities for attack and defense. Biologists have since used Huxley's metaphor in well over a hundred thousand articles. Julian Huxley's grandfather was Thomas Huxley, known as "Darwin's bulldog" for his strong advocacy of the theory of natural selection when it was a controversial new theory and had many powerful opponents.

4. It should be noted that the expression "survival of the fittest" was not coined by Darwin but by Herbert Spencer, another of Darwin's committed advocates (1864, 65). Spencer was later described as a "social Darwinist," and his ideas justified racial hierarchies and class inequities as a natural outcome of innate abilities.

5. Later, I remembered laughing over footage of some lion cubs playing together after hearing Attenborough say something like "while this looks like 'play,' these cubs are actually honing the skills needed for their survival." Somehow, something seemed awry in this harsh narrative of life with "no room" for play, but I could not quite describe my skepticism.

6. Sapp 2004.

7. Margulis 1998.

8. Trappe 2005.

9. McFall-Ngai et al. 2013.

10. Melin 1925.

11. Hiromoto 1963.

12. K. Hyde et al. 1998.

13. Understanding Covid-19's own particular liveliness is key to understanding it. Researchers were surprised to see that the virus is detectable for up to three hours in aerosols, four hours on copper, twenty-four hours on cardboard, and seventy-two hours on plastic and stainless steel (Van Doremalen et al. 2020). It also interacts with young and old human bodies differently from how other viruses do, being more virulent toward elders as compared to the 1919 epidemic, which mainly attacked the young. Covid-19 likely originated in other species (perhaps horseshoe bats from Yunnan), which is a reminder of the importance of our particular bodies and kinship

with fellow mammals, which allows the virus to jump between different species (Murdoch and French 2020).

14. Heckman et al. 2001.

15. Z. Li et al. 2016.

16. Gadd 2007.

17. Jongmans et al. 1977. The mycologist Shannon Berch told me that some details were known by Cromack et al. in 1979.

18. Jongmans et al. 1997.

19. Jongmans et al. 1997.

20. Frazer 2015.

21. Loron et al. 2019.

22. Heckman et al. 2001.

23. Kenrick and Strullu-Derrien 2014.

24. Söderström and Read 1987.

25. A number of these prehistoric plants have become extinct, but there are still some conifers, cycads, and lichens that have this quality. While some parts of these plants (such as leaves) undergo senescence, their entire bodies don't necessarily have a fixed or limited lifetime. Other trees, such as aspens, can live for vast periods of time by another method: propagating by clonal reproduction from root sprouts. Although individual aspen trees might live only a century or so, one colony is estimated to have existed for eighty thousand years (Finch 1994). Conventional animal models of life often assume that from the moment of birth, all organisms move through stages of life from youth to old age and toward death.

26. One of the few analogies to coal and petroleum—where the biological became mineral— is limestone. In certain locations, many millions of microscopic ocean animals (such as foraminifera) and plants (such as algae) gathered calcium carbonate from seawater. After they died, their bodies drifted down to the sea floor. In places where they were abundant, the sea floor became an underwater cemetery, and over vast amounts of time and pressure, their bodies compressed into limestone. Such minerals were locked away for long periods of time, perhaps hundreds of millions of years. The Yucatan Peninsula in Mexico, near where the last massive meteorite struck, is made entirely of limestone that was once the sea floor. Now that this landscape has become terrestrial, these ancient bodies are being broken down by and consumed by plants, fungi, and animals, reentering the calcium cycle and building new bodies.

27. Proponents of this theory include Cleal and Thomas (2005); Ward et al. (2006); Wilkinson (2006); and Tyler Volk (2007), and challengers include Nelsen et al. (2016). No matter the relative role of fungi in stopping fossil-fuel production, the role of fungi as the hidden supportive player that, through its mycorrhizal connections, helped make abundant plant life possible has rarely been told. Thus, in some ways, fungi made terrestrial plants possible and enabled the age of fossil-fuel-making, even if it turns out they weren't the major factor ending the fossil-fuel-making era.

28. Shae 2018; Vajda and McLoughlin 2004.

29. Large cold-blooded terrestrial dinosaurs relied on sunlight and warmth to be able to move actively, and with the dying-off of plants, herbivores suffered first, followed by the carnivores who fed on them. Smaller warm-blooded mammals, however, were not as dependent as reptiles on the sun to foster their metabolism and movement, as I will soon explain.

30. Biello 2011.

31. Dunn 2011.

32. Greenberg and Palen 2019.

33. Krings et al. 2007.

34. Strobel and Daisy 2003; Stone, Bacon, et al. 2000; Stone, Polishook, et al. 2004.

35. Koivusaari et al. (2019) argue, "Early papers hypothesized that all fungal endophytes are saprophytes or parasites, which slowly colonize the living leaf tissues and gain a head start over soil-dwelling saprophytes that invade the fallen, dead leaves." Perhaps a pathogenic bias against fungi played a role in the initial assumptions.

36. Thanks to Willoughby Arevalo for this expression.

37. This story is well known, but I'll provide a brief recap. In 1928, Alexander Fleming was studying staphylococci, a bacterium that causes serious staph infections. An untidy researcher, one of his petri dishes of staphylococci became contaminated, but before he threw it away, he noticed that the bacteria were dead. He cultured the contaminant, found it was a mold called *Penicillium* that was highly lethal to many kinds of bacteria. First named "mold juice," it was later deemed "penicillin," the basis for the antibiotic revolution, a story told with fascinating insights by Hannah Landecker (2016).

38. Fungal activity, for both plants and animals, turns what I call food into Food. I use the lowercase term to refer to its *potentiality* as nutrition and the uppercase term to refer to its *actuality* as nutrition, its absorption and incorporation into building tissue or creating energy. In the soil, for example, many forms of nitrogen are bound up with proteins that plant roots cannot access. Only after fungi or other microbes break up the proteins (food) into constituent parts are plant roots able to ingest them (Food). Likewise, rocks are food, and minerals are Food.

39. Dunn 2018.

40. Kendrick 2011.

41. Zhu et al. 2008.

42. Armstrong-James et al. 2014.

43. McFall-Ngai et al. 2013.

44. Boer et al. 2005.

45. Corrales et al. 2016.

Chapter Two. Everyday Fungal World-Making

1. Prestholdt 2018.

2. Zimmerman 2017.

3. Kimmerer 2013, 49.

4. TallBear 2011; Todd 2016.

5. In the meanwhile, they also coined the term "magic mushrooms" after being among the first Westerners to try them in Mexico. In 1957, Wasson wrote a hugely influential story in *Life*, one of America's top-selling magazines, that unleashed a wave of curiosity and experimentation. Timothy Leary, a Harvard professor, tried these magic mushrooms and wanted more, moving on to LSD to create "shifts in consciousness." Leary was a colleague of one of my relatives, Charles Slack, and I heard about their many experiments with LSD, including with priests and

prisoners. Leary remains an important figure of the 1960s counterculture movement, but recognition of Wasson's impact has faded, except among the "mushroom crowd."

6. Morris 1988.

7. On the other hand, this attitude is certainly not hegemonic, as England has produced a number of British mycologists who were fascinated with fungal abilities, such as Cecil Ingold, and Elsie Maud Wakefield, who I previously mentioned (Money 2016).

8. Huttner-Koros 2015.

9. Rolfe and Rolfe (1925) 2014, 1.

10. Berkeley was a pioneering British lichen expert, also called a cryptogamist. The term literally means "hidden marriage," as researchers could not understand how they reproduced, unlike flowering plants with more easily seen "male" and "female" parts. Linnean taxonomy placed fungi into the category of "cryptogams" along with ferns, moss-like organisms, and algae.

11. Berkeley 1857, 241, cited in Pouliot 2016, 4.

12. Hay 1887, 6.

13. Britt Bunyard, personal communication 2020.

14. Krebs and Elwood 2008, 153.

15. Trappe 2005, 277.

16. I have been curious to hear what happened in places like the Soviet Union, which had more complicated relations to neo-Darwinism. After a boom in Soviet mycorrhizal studies from approximately 1948 to 1963, this research went into a deep decline (Summerbell 2005).

17. Bonfante 2018.

18. Trappe 1977.

19. Varma, Prasad, and Tuteja 2017.

20. Ryan 2002, 24.

21. Tsing and Satsuka 2008.

22. Chrulew 2020, 142.

23. Dutton 2012, 92. Such mechanistic language is nearly hegemonic. Actions are typically assessed in terms of the "causal mechanism," a notion clearly part of a mechanistic paradigm. Such language presupposes a linear and predictable form of "stimulus and response" relationship. Although some predictable and causal relationships exist, I suggest that interspecies relationships are much more diverse, and such mechanistic models might miss this diversity. The notion of mechanism shapes the experimental context, often isolating an organism (such as an animal, a plant, a fungus) as an object of study in a cage where it is subjected to a single stimulus. In such a context, the possible reactions of the organism are greatly restricted compared to outside of the lab, where the organism lives in a complex world, full of multiple organisms, and where circumstances allow a wide range of possibilities. Thus, the stimulus-response model shapes the experiment context, which shapes the results in ways that reconfirm the initial model.

24. Riskin 2016, 2.

25. This is related, in some ways, to the notion of "environmental engineering," which I discuss in the conclusion. I anticipate that niche construction theory will become more important over time. Here is one promotional description for the book *Niche Construction: The Neglected Process in Evolution* (by F. John Odling-Smee, Kevin N. Laland, and Marcus W. Feldman, 2003) from the website of its publisher, Princeton University Press: "The seemingly

innocent observation that the activities of organisms bring about changes in environments is so obvious that it seems an unlikely focus for a new line of thinking about evolution. Yet niche construction—as this process of organism-driven environmental modification is known—has hidden complexities. By transforming biotic and abiotic sources of natural selection in external environments, niche construction generates feedback in evolution on a scale hitherto underestimated—and in a manner that transforms the evolutionary dynamic. It also plays a critical role in ecology, supporting ecosystem engineering and influencing the flow of energy and nutrients through ecosystems. Despite this, niche construction has been given short shrift in theoretical biology, in part because it cannot be fully understood within the framework of standard evolutionary theory" (https://press.princeton.edu/books/paperback/9780691044378/niche-construction). The idea that organisms' shaping of the environment creates a powerful evolutionary force is a provocative assertion (Laland 2004; Turner 2009), in turn challenged, in part, by Richard Dawkins (2004).

26. Turner 2009.

27. This is not exactly flipped, however, as animals do not dictate the environment and the environment does not adapt evolutionarily.

28. Some of the few proponents of such a dialectical view are the Harvard-based, Marxist-inflected biologists Stephen Jay Gould, Richard Levins, and Richard Lewontin. They largely focus not on the problem of active environment (as possessing a monopoly on what could be called "evolutionary agency") and nor at all on the organism (as I have described here), but on what they see as "genetic predeterminism," an overemphasis on the gene as the driver of life.

29. Although some imply that an active niche construction approach implies a form of intentionality in making a place better for an organism's own existence, a world-making approach does not necessarily imply intentionality. It is more interested in how organisms' activities shape the world around themselves.

30. Chamovitz 2012.

31. Cheplick 2015, 151.

32. Growing is often regarded as automatic, as a predictable unfolding, like the ticking of a watch hand. Movement, on the other hand, is often seen as intentional and dynamic, an action responding to a specific situation such as chasing prey. My daughter Logan Hathaway-Williams helped me think through these distinctions. Plants, especially trees that grow over a long period of time, seem to exhibit a kind of slow response to sun, wind and soil, more of a passive adaptation rather than a dynamic engagement. In the dominant narrative, fungi seem even less influenced by their surroundings; they seem to just pop up from the earth, self-formed, and are a relatively consistent shape and size, unlike the way trees can be sculpted by the wind over time.

33. Schilder 2020. Admittedly, many textbooks were written decades ago, before new findings in symbiotic relations were well known. However, the 2015 open-source textbook *Concepts of Biology* (Canadian edition; Molnarand Jane Gair 2015) never mentions mycorrhizae. It mentions fungi only as pathogens, with one brief mention that they create antibiotics and one paragraph on their life cycle. One of the most popular textbooks, *Campbell Biology* (Urry et al. 2016) has nineteen pages on fungi, yet this pales in comparison to long separate chapters on plants (145 pages) and animals (291 pages). *Campbell Biology* does mention mycorrhizae. We could have more hope for David Beerling's wonderful book, whose title promises an active, potentially

world-making perspective: *The Emerald Planet: How Plants Changed Earth's History* (2017). Beerling, however, neglects the role of fungi, mentioning them only four times and only then as rotters, ignoring their critical role in enabling trees' survival and flourishing through mycorrhizae (2017, 42).

34. Gagliano 2017, e1288333.

35. Wright and Jones 2006, 204.

36. Sutton 1996.

37. Frazer 2011.

38. Skerratt et al. 2007, 125.

39. Stamets 2005.

40. Schwartz 2013.

41. Sally Smith and Smith 2011; Leake et al. 2004; Jakobson 1992, cited in Hoysted et al. 2018.

42. Jennings 1995.

43. Jennings 1995.

44. Jennings 1995.

45. Archer and Wood 1995.

46. Riskin 2016.

47. Riskin 2016, 3.

48. Myers 2015; Pollan 2013.

49. Sheldrake 2020.

50. Ingold 2015; Griffiths 2015; Matsutake Worlds Research Group 2009a and b.

51. Broucek 2014.

Chapter Three. Umwelt: The Sensorial Experience and Interpretation of the Lively World

1. Nagel 1974.

2. See Uexküll 1909; Uexküll 2010.

3. Francis 2020. These include hermoception (sensing heat), nociception (pain), equilibrioception (balance), and proprioception (body awareness).

4. Uexküll was interested in plants and fungi to some extent, but he believed that because they don't have a central nervous system, they don't experience signs in as rich a way as animals do (Tønnessen, Maran, and Sharov 2018). However, recent studies of animals without a central nervous system, such as octopi, are expanding awareness of their vast capacities beyond what was previously assumed (Godfrey-Smith 2016; Montgomery 2015). My hope is that rather than understand the octopus as a form of "invertebrate exceptionalism"—such as how they are now frequently described as "*the* intelligent invertebrate" (Mather et al. 2013)—researchers will, instead, spend more effort to discover the capacities of others "without backbones," often wrongly assumed to be "simple animals" (Matheson 2001, 1).

5. Some researchers, such as Konrad Lorenz, were inspired by Uexküll in the beginning of their careers but later spoke out against him after the rise of neo-Darwinism, because Uexküll was labeled a "vitalist" critic of Darwin's theories (Brentari 2009).

6. This resulted in the unlikely discovery by Karl von Frisch that bees dance to communicate either the direction or the distance to nectar, or the quality of nectar available there, or possible

nesting sites for a swarm, and in order to recruit other workers to these sites. An observer well versed in this language can watch the dance and learn exactly where the bees are foraging. This example shows that while a utopian twinning may be unreachable in Nagel's terms (1974), that sustained and careful attention, along with a curiosity motivated by a belief in other organisms' complexity, can lead to much better understanding.

7. Ironically, even though Uexküll was skeptical of Darwin's theory of evolution, Darwin was also paying close attention to worms and their abilities at the same time (1892). They carried out some similar experiments such as watching how worms react to various scents and how they invent strategies for dragging different kinds of tree leaves into their underground burrows. Worms seem to classify leaves into two shapes and learn which end to pull down first into their tunnels, the stem or the tip.

8. Ingold 2000; see also Magnus 2014.

9. Hayward 2010.

10. Maran et al. 2012, 12.

11. This is also true in terms of British forms of measurement, where size was determined by the "human body," and in particular the body of one large man (King Henry): in the twelfth century, a yard was fixed as the distance from his nose to the thumb of his outstretched arm, turning his body into a national standard. Likewise, many cultures came up with an approximate foot, based on the human foot (albeit a particularly large one). The mile was a by-product of Roman imperial expansion, based on a thousand military paces.

12. Chamovitz 2012.

13. Chamovitz 2012, 10.

14. Some similar problems occur with terms such as "learning" and "decide." As reported to Michael Pollan (2013), one scientist who attended a conference that discussed "plant learning" argued "that 'learning' implied a brain and should be reserved for animals" (98). As Pollan reports, the scientist said: "'Animals can exhibit learning, but plants evolve adaptations'. . . . At lunch, I sat with a Russian scientist who was equally dismissive. 'It's not learning,' he said. 'So there's nothing to discuss'" (98). Pollan also quoted plant physiologist Lincoln Taiz as stating, "The verb 'decide' is inappropriate in a plant context. It implies free will" (104). I would question that decisions are based on "free will" and the assumption that only animals can learn. It seems that for some scientists, learning is conceptually limited to animals and it is a waste of time to study learning in non-animals, a narrowness of definition that is debatable.

15. Chamovitz 2012, 48.

16. Sterne 2003, 11.

17. Roosth 2018, 110.

18. Appel and Cocroft 2014. The widespread interest in this paper is underappreciated by official citations. In April 2020, it was cited by eighty-five peer-reviewed papers but accessed eighty-six thousand times on the journal's site. It speaks to a hunger for more knowledge on more-than-human powers.

19. Veits et al. 2019.

20. Gagliano 2017.

21. Corn seedlings, for example, make a clicking sound.

22. Aggio et al. 2012.

23. Horowitz 2010.

24. Horowitz 2010, 255.

25. In other words, assuming that a person is visually oriented and doesn't have as rich a sense of layered smell as a dog, what they see in the moment reveals only what happened then, rather than what is happening up ahead or happened earlier. Although we can imagine the future and the past in our minds, we don't tend to see it all in one moment as one can with smell.

26. Magnus 2014.

27. Healy et al. 2013.

28. Gaycken 2012, 54.

29. Wohlleben 2015.

30. Chamovitz 2012.

31. Z. Yu and Fischer 2019.

32. Rodriguez-Romero et al. 2010.

33. Lorimer 2015, 41. In the interest of not assuming human exceptionalism, it is likely that humans may communicate more through pheromones than we currently realize (Doty 2014).

34. In the neighboring discipline of biological anthropology, my peers study bodies and perception to compare *Homo sapiens* to our primate cousins and explore how our bodies adapted to different conditions as we spread out of Africa.

35. Yong 2015.

36. Yong 2015. Although there are a number of English words used to describe one's feelings about a smell (comforting, intoxicating), its intensity (strong, powerful), or its similarity to something (citrusy, garlicky), there are very few words that describe its specific qualities, like the way we label specific colors.

37. Classen 1993.

38. Kant 1798, 50.

39. Horowitz 2016.

40. Kish 2009; Nagel 1974.

41. See, for example Laland and Galef (2009) and debates such as Laland and Janik (2006) and replies by Krützen et al. (2007).

42. Agamben 2004, 46.

43. Only a few social scientists, such as Deleuze and Guattari, have gone beyond the tick example. They explore the orchid and the wasp, showing how their bodies and their sensorial apparatus arose out of a long evolutionary dance. This contributes to their notion of a being's "becoming," which arises through many encounters with others (Deleuze 1987).

44. Sebok 2009, 223.

45. Hallé (1999) 2011.

46. Witzany and Baluška 2012; Witzany 2012.

47. Witzany 2006, 169.

48. Witzany 2012, 1.

49. Witzany 2006, 170.

50. Hughes 1990.

51. Felton and Tumlinson 2008.

52. McCormick et al. 2012.

53. Pálsson 1994; Ingold 2000.

54. Barua and Sinha 2019.

55. Walsh et al. 2011.

56. Witzany 2006, 170.

57. Macfarlane 2016.

58. E. Newman 1988, quoted in Macfarlane 2016.

59. Simard et al. 1997.

60. Dawkins 1976.

61. Luginbuehl and Oldroyd 2017, R953.

62. Robinson and Fitter 1999.

63. Likewise, as I mentioned in Chapter 2, Rob Dunn suggests that while almost all cases of insect-fungi relations that seem like domestication are described as "insects farming fungus," one could easily flip the script and see "fungi [as] farming insects," getting animals to take care of them and expand their range (2012). This points to a dichotomous conceptual model where one party is seen as active and the other as passive, which is increasingly challenged by new ways of understanding domestication as a coevolutionary mutualistic relationship compared to an active agent domesticating a passive agent (Zeder 2015). Regardless, humans were about sixty million years late in farming fungi compared to three different insect lineages (beetles, ants, and termites), which gained this ability independently of each other (Mueller and Gerardo 2002).

64. Wohlleben 2015.

65. This is unless, perhaps, their mycelia are surrounded by frozen soil or desiccated and in a state of suspended animation.

66. Ingold 2011.

67. Maser et al. 2008. Although some disregard such communication as "mere chemicals," unlike the sophistication of human speech, one can also understand that chemicals from our bodies inflect human relations as well, shaping the dynamics of bonding with infants, sexual desire, sense of fear, and so forth.

68. Pyare and Longland 2002.

69. Hochberg et al. 2003.

70. This is technically in the same genus as *Cordyceps* but has a different Latin name. This is the anamorph stage (asexually reproducing), whereas *Cordyceps* is the teleomorph stage (sexually reproducing). Thus, many fungi contradict the typical scientific rules for naming whereby one organism has only one Latin name. Many fungi have radically different life stages (think of a caterpillar and a butterfly for an insect analogy). In many cases, scientists were not able to follow them through their stages, as the second stage usually needs a certain partner organism to come into being, and such partnerships, for interesting reasons, do not seem to happen often under the watchful eyes of a lab scientist. The anamorph lives around the world, but the teleomorph lives only in Asia. In the case of this fungus, it would be as though the caterpillar could reproduce without ever metamorphosing into a butterfly, or there were self-cloning caterpillars throughout the world, but only in Asia could they turn into butterflies that had sex.

71. George et al. 2013.

72. Although often feared and destroyed by farmers in the United States and Canada, in Mexico some farmers try to spread the fungus, for it is delicious and fetches a far higher price from consumers than the corn itself.

73. Jennersten 1988.

74. Sivinski 1981.

75. This is just one example out of many where scientists still apply plant-based terms to the fungal world that are literally untrue, for the term *pollinator* means to carry pollen. Here are three other examples, still commonly in use: "fruiting body," for which a better term is "sporo-carp" (meaning "spore-bearing body"); "vegetative growth," which would be better described as mycelia; and "saprophyte" as a fungus that eats the dead, but the Latin suffix "phyte" means plant, so a better alternative term would be "saprotroph." Water-based fungi are referred to as "phytoplankton" but should be called "macroplankton." There are also examples that show the persistence of a "two kingdom model": doctors refer to bacteria as our "intestinal flora," and some water-based spores are called zoospores (animal spores) because they can swim (assumed to be an animal-only trait), but some fungi and plants have swimming sperm or spores.

76. Witzany 2012.

77. Again, as *Nature* magazine had coined it in 1997; see also Witzany 2012.

78. Choy 2018.

79. George Herbert Lewes was the first known person to apply the term "anthropomor-phism" to animals. Lewes believed the mollusks he researched "had only rudimentary sensitiv-ity to light" and said, "We speak with large latitude of anthropomorphism when we speak of the 'vision' of these animals. . . . Molluscan vision is not human vision; nor in accurate language is it vision at all" (1860, cited in Wynne 2007, 126). Although I applaud Lewes for not conflating human and mollusk vision, he uses human sight as the standard for what will count as "vision." He assumes that, rather than inquires whether, mollusks have rudimentary visual capacities, and he uses an unspecified definition of vision to dismiss the mollusk. Uexküll offers a far richer approach by refusing to use humans as the standard. Instead, he was interested in how other organisms possess senses potentially far more developed or just plain different from our own.

80. For instance, many scientists are loath to write that organisms can "decide" to take any course of action, as this verb is taken to be an anthropomorphism (Riskin 2016), and some think this ability is available only to animals. Others assert that decision-making "requires free will," a quality assumed to be possessed only by humans (Pollan 2013). In many cases, scientists assert that a certain characteristic is human-only rather than testing that claim themselves or carefully consulting the published literature. Although many avoid anthropomorphism in their publica-tions, they substitute another form of morphism: mechanicomorphism (Crist [1999] 2010; De Waal 1999), which describes organisms as machines.

81. I had initially assumed that biosemiotics was my key to uncovering the realm of active multispecies world-making; I now realize that it is one intriguing and helpful approach but cannot address everything I am curious about. Moreover, certain neo-Uexküllian approaches and styles of biosemiotic analysis tend to create a disembodied sense of the organism, and to understand organisms in mechanicomorphic terms that strip them of the authorship of their own actions, turning them from subjects into objects. Biosemiotics tends to privilege informa-tion above all else, while perception is sometimes depicted as a schematic of how a machine works as electricity flows through sensors, activating some triggers and directing its path with feedback loops. The object that moves along this path is information. Some of the principles of biosemiotics, I was told, were also influential in creating the field of cybernetics (machine learn-

ing, which brought us computers and "artificial intelligence"), and I could see the family resemblance. Rather than focus on information, I advocate focusing on embodied movement through and in relationship to a messy world full of dynamic encounters.

Chapter Four. Matsutake's Journeys

1. The landscape described in the epigraph to this chapter as the Third Pole, squeezed like an accordion and nourished by glacial melt, is also the birthplace of some of the largest and most important rivers in Asia and indeed on the planet. The eastern-flowing Yangtze (Jinsha) River stays within China; the southern-flowing Mekong (Lancang) River, the Salween (Nujiang) River, and the Irrawaddy (Dulongjiang) leave China to travel into Burma, Thailand, Vietnam, Laos, and Cambodia. Together, this relatively small area plays a major role in shaping a vast region.

2. Anna Tsing provides an intriguing account of matsutake's adventures that focuses on its globe-trotting movements and explores how scientists try to learn about these journeys (2015, 218–20).

3. Patowary 2017.

4. For readers elsewhere, here are some equivalent distances. For the United States: from Maine to Florida; for Canada: from Winnipeg to Montreal; for Europe: from Edinburgh to Rome; and for Africa: from Khartoum to Nairobi.

5. Y. Guo et al. 2017.

6. Y. Wang et al. 2017.

7. In Yunnan, these forests are mainly pine (*Pinus densata*, *P. yunnanensis*, and *P. armandii*) and oak (*Quercus semicarpifolia*, *Q. aquifoliodes*, *Q. senescens*, *Q. panosa*, *Q. gillilana*, and *Q. spinnosa*) (Y. Wang et al. 2017).

8. Indeed, as mentioned in chapter 1, mycorrhizal relations were likely critical to the movement of plants from water to land hundreds of millions of years ago. Fungi enabled small plants to evolve into massive trees, which have substantial requirements for flows of water and minerals.

9. Berch et al. 2016. For example, the highly toxic "death cap" (*Amanita phalloides*) received a lot of attention as a new arrival to British Columbia, Canada. For a while, it seemed to connect only with introduced street trees, but Shannon Berch told me that she saw it fruiting under a native Garry oak in 2015. Death caps likely traveled from Europe to California during the 1930s, hitchhiking in potted trees from the nursery trade.

10. These are increasingly understood to be bioelectric as well (Volkov and Shtessel 2020).

11. Arora 1986, 191. David Arora is a prominent mycologist, well known for his best-selling and bible-thick identification manual called *Mushrooms Demystified* (1986).

12. Not all cultures appreciate this smell. A closely related matsutake species found in Europe, which is now shipped to Japan as part of the species complex, was given the Latin species name of *naseolum*: "that which makes one nauseous." In Japan, the fact that some non-Japanese have a distinct dislike of the smell reinforces a sense of cultural distinction.

13. I found that many North American scientists assumed such odors were defensive in nature, what some call an "antifeedant," and they used slugs to test this hypothesis. I was intrigued that

some Japanese scientists researched how odors work as insect attractants, which reconfigures the interaction differently: rather than a mushroom defending itself, it is attracting insects (likely to act as its dispersal agents, its spore vectors). On a related note, some mycologists in Canada studied the preferences of several slug species. Their study foregrounds desire and attraction, not defense (Voitk et al. 2012).

14. Glass et al. 2004.

15. These insights are from Tom Volk: https://botit.botany.wisc.edu/toms_fungi/feb2000.html.

16. Some asexual fungi can make spores (Watling 1980).

17. Those curious to learn more about fungal sex lives can also consult Tsing 2015 (216–18, 222–26) and, for a "magical mystery tour," listen to Willoughby Arevalo's version at https://www.youtube.com/watch?v=ldOR1rUkNOI&t=2141s and https://www.youtube.com/watch?v=bbKiIbAprNQ&t=1703s.

18. As Anna Tsing describes, in Oregon, especially when a light snow covers the ground, mushroom hunters are sometimes alerted to the presence of matsutake by trails of blood (2015). Elk's passion for matsutake means that some dig down into the sharp volcanic soil, following the mushroom to its base and cutting their delicate noses in the process. If the weather is too cold and the matsutake freeze solid, this can spell the end of the season, but with just the right conditions, the elk become a kind of beacon to alert the hunter to prolific patches. Even though an elk might eat all the matsutake in a patch, this patch may produce more mushrooms in another wave, a "flush."

19. I have seen where birds have pecked matsutake, but I have not yet discovered any reports of particular species of birds, let alone reptiles or amphibians, eating them. Many reptiles, especially turtles, enjoy eating fungi. Ornithologists who study bird diets typically do not look for traces of fungi (Elliott, Bower, and Vernes 2019). In terms of amphibians, it is well known that fungi are devastating frog and salamander populations. Green frog tadpoles can turn the tables and eat "fungi," but the specific species was unspecified (Jenssen 1967).

20. Lefevre 2003, 2.

21. Qian et. al 1991.

22. Like with so many other species, global trade might influence matsutake's movements and connections. Perhaps the trillions of spores from shipments of mushrooms coming into Japan are flying around and one or more has germinated, made relations to trees, and even mated with existing ones. As far as I know, no study exploring this possibility has been carried out.

23. K. Li 2004.

24. B. Yang 2004.

25. Logging was most intense where forests grew along gentle rivers upstream of major cities. Especially after periodic fires burned cities down, loggers cut massive stands of trees, which they floated downstream (Menzies 1994). Yunnan's capital of Kunming was far from the Yangtze, the closest major river. This river is typically inhospitable to floating timbers, with its steep banks, fast current, and white water. In the few places where large forests grew along the river's bank, they were cut out, but in most of Yunnan, where more than three quarters of the land is mountainous, this landscape made it difficult to transport timber before vehicle roads were built.

26. Yunnan was well known for its mines, some dating back a millennium and long since tapped out, while others are still going strong, producing vast quantities of copper and tin (Kim 2018). Mines burned vast quantities of trees to process the ore.

27. Lashio was at the end of a rail connection from the port of Rangoon, the capital of British Burma. Although it is frequently implied that the road was built from scratch, it often followed one of the paths of the ancient "Southern Silk Road" through the rugged terrain. Soon after it was finished, however, Japanese troops expanded into Burma and cut off the road for several years, marching into Yunnan, sometimes facing up to a hundred thousand Chinese troops.

28. In the early 2000s, I met Tibetan matsutake hunters who discovered some of these plane wrecks, and they showed me some tools they had recovered from them.

29. MacMurray 2003.

30. Pierson 1958.

31. Yamanaka, Yamada, and Furukawa 2020, 49.

32. Many people have asked me if matsutake, being the most valuable mushroom in Japan, is also the most loved in terms of flavor. The answer to this is interesting, as there is a common saying in Japan: "Nioi matsutake, aji shimeji," which means "matsutake for the smell, shimeji for the taste." When I tell people this expression and say that shimeji are easily cultivated and have become cheap, they often look puzzled: if the best-tasting mushroom is cheap, why spend a lot of money on a mushroom that *just* smells good? I think, however, that we cannot say it just smells good; this smell is part of its quintessential nature that is utterly distinctive, and it carries with it not only an intensely seasonal quality, but the weight of Japanese history and identity. More than once I heard Japanese dealers or consumers say that only Japanese people love matsutake and that this idiosyncratic love is evidence of their unique and distinctive identity. Others commented that while they love the smell of matsutake in the market, they actually don't know how it tastes, as they consider it "too expensive" to buy!

33. Wilkening 2004. Another explanation for Japan's forest decline points to Japan's rural exodus: rural people had regularly cut hardwood trees for fuelwood and raked away forest duff for animal bedding. Ironically, these two activities that North Americans tend to think of as damaging to forests actually created excellent matsutake habitat, as the mushrooms couldn't grow with hardwoods but did grow with the pines that started to grow in cut over forests. Eventually, however, these pine trees, already aging and weakened by acid rain, also suffered from an introduced microscopic, worm-like pest, a nematode, which further diminished production (Saito and Mitsumata 2008; Amend et al. 2010; Faier for the Matsutake Worlds Research Group 2011).

34. Wilkening 2004.

35. Muldavin 2000.

36. Bracey 1990, quoted in Redhead 1997. These figures are approximate. Unlike manufactured goods, such as cars, that are produced in large factories and easily quantified, quantities of wild products such as matsutake can be traded below the radar of state agencies, which is often reflected in the fact that China's export figures and Japan's import figures for matsutake rarely match. Only in 2000 did the Yunnan government begin to actively aggregate data on matsutake production; previously, such data was generated by different government sectors and offices, and mushroom species were often combined or mixed with other nontimber forest products such as bamboo shoots.

37. In the United States, Japanese Americans hunted matsutake in places like Oregon's Mount Hood by the 1930s, and Japanese Canadians discovered matsutake at World War II internment camps (Zeller and Togashi 1934; Shiho Satsuka, 2011, personal communication). Although some mushrooms were shipped back to Japan, they decomposed long before reaching their destination.

38. Redhead 1997.

39. Kinugawa and Goto 1978.

40. Most scholars concur that before the 1980s, matsutake generated little interest or worth in Southwest China, for domestic use or export. Daniel Winkler found mention of one exception in the forestry history of Ganzi Tibetan Autonomous Prefecture, which claims that "between 1909 and 1912, ten tons of matsutake at a total value of four hundred kilograms of silver were exported from Kangding [in present-day Sichuan]" (2008, 9). To my knowledge, this is the only citation on China's matsutake trade before the 1970s. I am not completely convinced, however, that this *song rong* is indeed *Tricholoma matsutake*, as common names can sometimes refer to several different species. We have no further details, such as where these mushrooms were sent, but the shipments would have likely been dried.

41. Tominaga, Arai, and Ito 1981.

42. Rowe 1997.

43. Yet this is not always fixed for in Japan in 2004, Chinese matsutake fetched a higher price than Canadian ones (Gong and Wang 2004).

44. X. Yang et al. 2006.

45. Amend 2009.

46. In 2008, the Chinese government petitioned for matsutake to be deemed an endangered species. This move surprised many and generated a number of critics. It led to a series of laws that restrict international sales of small matsutake, but Kunming dealers created a domestic market, at first, for "hot pot." Demand has continued to increase (Faye 2016; Junqian Xu 2017). In Oregon, such small mushrooms are called "babies" and are usually classified into the number 3 grade, thus fetching a relatively low price compared to those gathered a day or two later (Tsing 2015).

47. Like many scientific claims, this one is also challenged (see Sha et al. 2007 for one study).

Chapter Five. The Yi and the Matsutake

1. Fiskesjö 1999.

2. Bradley 2001.

3. Yi groups were likely prominent in the Nanzhao Kingdom, which flourished during the eighth century and was powerful enough to fight the Tibetan Empire to the west and Tang China to the east. By the twentieth century, one Yi leader, Long Yun (who played an important role in the toppling of the Qing dynasty), became Yunnan's governor from 1927 to 1945 until removed by the Nationalists. After the new PRC government came to power in 1949, Long regained his position until his death in 1962.

Despite Long Yun's support from the new government, many Yi did not immediately embrace changes. The new government tried to eradicate the opium that was key to their economy. Anti-opium campaigns were likely a major reason behind a series of Yi rebellions that lasted for almost two years, from 1955 to 1957. The Yi were well armed: in the neighboring Liangshan area of southern Sichuan, some officials estimated that locals possessed more than one hundred thousand guns (Zhou 1999, 152). The PRC's first opium poppy destruction campaigns started in 1955, but after Yi killed a dozen cadres, the campaign was quickly disbanded. As I describe in the next chapter, the PRC army was also kept busy by two massive campaigns: suppressing a series of Tibetan rebellions from 1956 until 1959, and supporting the Vietnamese. Supplies and trainers were sent starting in 1950, thousands of railway workers in 1954, and over two hundred million yuan in 1955, first against the French army and later against the Americans.

As some older Yi remembered, Long Yun led Yunnan for almost two decades, but there were still many places where opium dealers reigned. Yun was busy fighting tens of thousands of Japanese troops entering Yunnan from Burma. As one Yi gentleman, in between puffs of tobacco from his pipe, explained, "In the 1940s, there were still a number of local warlords who controlled their area. The Japanese were coming into Yunnan, and the KMT (Nationalist) soldiers were fighting them. Some landlords had lots of money from opium; they were rich." A century earlier the British had created the transnational opium trade, growing the crop in India and selling it against the will of the Qing state. Despite British efforts to dominate the trade and Qing efforts to repress it and ban domestic production, Southwest China surpassed India as the main source of the drug by the end of the nineteenth century. Why? Profits were vast; Qing rule was often precarious, with few Han officials in the region, in part because of their fears of rampant malaria and other diseases; and opium growers were well armed (Bello 2003). By 1957, the CCP largely accomplished what the Qing state failed to do almost 130 years ago: it wiped out opium. Yet seventy years later, rumors abound that people here are once again growing poppies.

4. Mueggler 2001.

5. Hathaway 2013.

6. While the Han have been eroticizing the Dai as a people of beautiful women and mysterious intrigue for decades (S. Hyde 2007), there is almost no equivalent for the Yi, who are largely seen by Han as backward mountain people.

7. In Nanhua, I was impressed that some of the Yi bosses talked about using the profits to support themselves as a culture. In talking to Anna Tsing about her fieldwork in the US Pacific Northwest, I did not hear about these kinds of equivalents, although there are some kinds of ethnic convergences around buying stations. One of the key differences in this part of Yunnan is that almost all the matsutake hunters I talked with lived in local villages, and their families had done for generations, whereas in the US many traveled far distances to pick.

8. In Japan, however, many if not most matsutake never become commodities, and a number are given to others as gifts. The Japanese are not alone. In Vancouver, two of my neighbors, an elderly Korean couple, told me about their joy in waking up in the fall and finding a basket of freshly picked matsutake right outside their door. When I lived in Michigan, many neighbors were crazy about picking morels. We gathered them by the basket, never selling a single one. After we dried them, they were more precious than anything from a store. Given away in small bags, or used to cook a meal for special friends, these mushrooms not only were delicious but possessed wonderful and ineffable qualities, especially for friends who were not morel hunters. It mattered that my partner had found them herself, using skills that many people did not have; they never became a commodity.

9. Butler 1988.

10. While some organisms decay with the assistance of eaters like bacteria and insects, others create their own autodigestion such as the shaggy mane (*Coprinus comatus*) mushroom, which spreads its spores by turning its spore-laden cap into liquid goo.

11. Y. Yang et al. 2007; S. Li et al. 2010; Su Kaimei, interview 2009.

12. Xue 2006.

Chapter Six. Tibetan Entanglements with Plants, Animals, and Fungi

1. Yeh 2000; He 2010.

2. D. Yu 2013.

3. Goldstein 1997.

4. J. Li 2016, 331.

5. D. Mortensen 2016.

6. D. Mortensen 2016.

7. See China Statistical Bureau figures for 2013, accessed March 31, 2016, http://www.tjcn .org, cited in Yeh and Coggins 2014, 8. As well, a few Hui, Pumi, Miao, Nu, and Dulong peoples live in the prefecture, each less than 1 percent of the registered population (Kolas 2007, 2).

8. Wu and Wu 2004, cited in Rhode et al. 2007, 205. To explore another analogy, think of the critical tripartite relationship between the free-ranging cow, the horse, and the cowboy, who all shaped each other. Likewise, there are no fishermen without fish, no farmers without plants that are able to be farmed, no herders without animals able to be herded.

9. Johnson 2018, 5.

10. F. Chen et al. 2015.

11. Liu et al. 2017.

12. Others contend that Tibetan barley was descended from a local wild barley (Laurent 2015).

13. Tashi et al. 2013.

14. D. Miller 1999.

15. These yaks are close enough to their wild ancestors that they interbreed and produce fertile offspring.

16. Rhode et al. 2007. A number of non-Tibetan groups raise yaks, and of these, nomadic herders have more intense everyday relations to yaks than permanently settled groups (Joshi et al. 2020). Yaks themselves do not stay settled, for unlike animals such as pigs (which often stay in a pen), yaks almost always travel to high-altitude summer grasslands.

17. Quoted in Peters-Golden 2012, 164.

18. Below this elevation, the best caravan animal is often the mule, which has been used by Muslim traders for centuries. Between the favored yak territory and the land of the mule is a zone where a new animal thrives, a cow-yak hybrid called *dzo* (males) and *dzomo* (female).

19. For more than a millennium, raising horses was a major part of the Tibetan economy. During the tenth century, Tibetans traded more than twenty thousand Tibetan warhorses a year with the Chinese court for vast quantities of tea. In 1735, the court stopped buying these horses,

radically transforming the horse market. Of course, Tibetans did not let this event end their quest for acquiring tea, soon trading with other partners. See the 2004 newsletter of the Silk Road Foundation, http://www.silkroadfoundation.org/newsletter/2004vol2num1/tea.htm.

20. Some scholars wish to emphasize animals' agency in terms of their transgression of human intentions (Philo and Wilbert 2000; Hribal 2007), but I would rather explore the broad range of animal action beyond a human center.

21. For scientists, the notion of nonhuman intentionality is often a taboo concept in constructing a biological model of the world that tries to scrupulously avoid any association with purpose or goals, concepts linked to religion. For example, some critics of the term "ecosystem engineering" (which describes the actions of organisms that influence their environment) reject the term in part because they think it implies intent on the animal's behalf (Wright and Jones 2006).

22. Jianchu Xu and Wilkes 2004.

23. Winkler 1998.

24. Jianchu Xu and Wilkes 2004.

25. B. Li 2002.

26. Tsing 2015.

27. Hathaway 2017.

28. As we learned earlier, trees alone do not enrich soils. Mycorrhizal fungi pull minerals and water up from deep underground, which feed leaves and branches. In turn, leaves and branches fall to the ground and are digested by fungi and bacteria, which enrich the soil with organic matter. Over time, this cycle makes hard soil soft, making the land more like a sponge that can absorb rainwater that might otherwise flow into the Yangtze River.

29. Northwest Yunnan's lack of industry could have been different. In nearby Panzhihua in Sichuan Province, Beijing created a massive industrial city in the 1960s based on their fear that the Soviets might bomb their eastern cities.

30. Brandt et al. 2012.

31. This was told to me by a local official. Susan McCarthy writes that in 2006, the average per capita rural income in Yunnan was US$281 (2011:18), and northwestern Yunnan tended to have lower income, so this seems possible.

32. Arora 2008, 284.

33. With the exception of the Mongolians (Yuan dynasty) and the Manchu (Qing dynasty).

34. Just three years later, Tibetans' charismatic leader, the Dalai Lama, fled China to India, where he set up a government with eighty thousand others. He motivates other countries to raise Tibetan issues during negotiations with China. Writings on Tibetan history tend to be polarized. For a sophisticated treatment of some of these dynamics, see Carole McGranahan (2005, 2010, 2019) and Charlene Makley (1997, 2007).

35. For example, a 1994 policy banned photos of the Dalai Lama and forbade any government worker from "engaging in religious practice or activities" (Barnett 1996, 56).

36. Junqian Xu 2017.

37. There are conflicting reports on China's poverty line (Glauben et al. 2012).

38. Hayes 2013.

Chapter Seven. Final Thoughts on Understanding Fungi and Others as World-Makers

1. Kimmerer 2003.

2. Shigo 1979.

3. Morell 2016.

4. Rayner 1997; Ingold 2015.

5. Bennett 2010.

6. Latour 1984.

7. Hustak and Myers 2012; Beer 2000; Riskin 2016; and Grosz 2011.

8. Tsing et al. 2017.

9. Smil 1980.

10. Zhong and Shi 2020, 15.

11. Heidegger 1995, 320–21.

12. Boyd 2012.

13. Ruru 2018.

14. Tsing 2012; Tsing 2015; Haraway 2003; Haraway (2007) 2013; Plumwood 2002a; Plumwood 2002b; Myers 2015; Todd 2014; Todd 2016; De Waal 2008; Rose 2011; Lien 2015; Lien and Pálsson 2019; van Dooren 2019; van Dooren and Rose 2016; Hird 2000; Hird 2004; Despret 2016.

15. Haraway 2016, 30.

16. Arevalo 2019; Steinhardt 2018.

17. McCoy 2016.

18. Khan et al. 2017.

19. Ittner et al. 2018.

20. Khan et al. 2017.

21. Lévi-Strauss 1963, 89.

22. Fang et al. 2018.

23. See, e.g., Hou et al. 2013; Yin et al. 2012.

24. Tsing and Satsuka 2008, 250.

25. Tsing 2015, 250.

26. Satsuka 2018.

27. Satsuka 2018, 90.

28. Bilger 2007.

BIBLIOGRAPHY

Adl, Sina M., Alastair G. B. Simpson, Christopher E. Lane, Julius Lukeš, David Bass, Samuel S. Bowser, Matthew W. Brown, Fabien Burki, Micah Dunthorn, and Vladimir Hampl. 2012. "The Revised Classification of Eukaryotes." *Journal of Eukaryotic Microbiology* 59, no. 5:429–514.

Agamben, Giorgio. 2004. *The Open: Man and Animal.* Palo Alto, CA: Stanford University Press.

Aggio, Raphael Bastos Mereschi, Victor Obolonkin, and Silas Granato Villas-Bôas. 2012. "Sonic Vibration Affects the Metabolism of Yeast Cells Growing in Liquid Culture: A Metabolomic Study." *Metabolomics* 8, no. 4:670–78.

Amend, Anthony S. 2009. "MycoDigest: Matsutake at the Roof of the World." *Mycena* 60, no. 7:1–8.

Amend, Anthony, Zhendong Fang, Cui Yi, and Will C. McClatchey. 2010. "Local Perceptions of Matsutake Mushroom Management, in NW Yunnan China." *Biological Conservation* 143, no. 1:165–72.

Appel, Heidi M., and R. B. Cocroft. 2014. "Plants Respond to Leaf Vibrations Caused by Insect Herbivore Chewing." *Oecologia* 175, no. 4:1257–66.

Archer, D. B., and D. A. Wood. 1995. "Fungal Exoenzymes." In *The Growing Fungus*, edited by Neil A. R. Gow and Geoffrey M. Gadd, 137–62. New York: Chapman and Hall.

Arevalo, Willoughby. 2019. *DIY Mushroom Cultivation: Growing Mushrooms at Home for Food, Medicine, and Soil.* Gabriola Island, BC: New Society.

Armstrong-James, D., G. Meintjes, and G. D. Brown. 2014. "A Neglected Epidemic: Fungal Infections in HIV/AIDS." *Trends in Microbiology* 22, no. 3:120–27.

Arora, David. 1986. *Mushrooms Demystified: A Comprehensive Guide to the Fleshy Fungi.* Berkeley, CA: Ten Speed.

———. 2008. "The Houses That Matsutake Built." *Economic Botany* 62, no. 3:278–90.

Bagemihl, Bruce. 1999. *Biological Exuberance: Animal Homosexuality and Natural Diversity.* New York: Macmillan.

Baldwin, Ian T., and Jack C. Schultz. 1983. "Rapid Changes in Tree Leaf Chemistry Induced by Damage: Evidence for Communication between Plants." *Science* 221, no. 4607:277–79.

Barnett, Robert. 1996. *Cutting off the Serpent's Head: Tightening Control in Tibet, 1994–1995.* New York: Human Rights Watch.

Bar-On, Yinon M., Rob Phillips, and Ron Milo. 2018. "The Biomass Distribution on Earth." *Proceedings of the National Academy of Sciences* 115, no. 25:6506–11.

Barua, Maan, and Anindya Sinha. 2019. "Animating the Urban: An Ethological and Geographical Conversation." *Social and Cultural Geography* 20, no. 8:1160–80.

Bavington, Dean. 2010. *Managed Annihilation: An Unnatural History of the Newfoundland Cod Collapse*. Vancouver: University of British Columbia Press.

Beer, Gillian. 2000. *Darwin's Plots: Evolutionary Narrative in Darwin, George Eliot and Nineteenth-Century Fiction*. Cambridge: Cambridge University Press.

Beerling, David. 2017. *The Emerald Planet: How Plants Changed Earth's History*. New York: Oxford University Press.

Bello, David. 2003. "The Venomous Course of Southwestern Opium: Qing Prohibition in Yunnan, Sichuan, and Guizhou in the Early Nineteenth Century." *Journal of Asian Studies* 62, no. 4:1109–42.

Benedict, Ruth. 1934. *Patterns of Culture*. Vol. 8. Boston: Houghton Mifflin.

Bennett, Jane. 2010. *Vibrant Matter: A Political Ecology of Things*. Durham, NC: Duke University Press.

Bennett, Jane, and Klaus Loenhart. 2011. "Vibrant Matter, Zero Landscape: Interview with Jane Bennett." *Eurozine / GAM Architecture Magazine* 7:1–7. https://www.eurozine.com/vibrant-matter-zero-landscape/.

Berch, Shannon M., Paul Kroeger, and Terrie Finston. 2016. "The Death Cap Mushroom (Amanita Phalloides) Moves to a Native Tree in Victoria, British Columbia." *Botany* 95, no. 4:435–40.

Berkeley, Rev. M. J. 1857. *Introduction to Cryptogamic Botany*. London: Baillaire.

Berlant, Lauren, and Michael Warner. 1998. "Sex in Public." *Critical Inquiry* 24, no. 2:547–66.

Besky, Sarah, and Jonathan Padwe. 2016. "Placing Plants in Territory." *Environment and Society* 7, no. 1:9–28.

Biello, David. 2011. "It's Official: Fungus Causes Bat-Killing White-Nose Syndrome." *Scientific American*, October 26. https://www.scientificamerican.com/article/fungus-causes-bat-killing-white-nose-syndrome/.

Bilger, Burkhard. 2007. "The Mushroom Hunters." *New Yorker*, August 13. http://newyorker.com/magazine/2007/08/20/the-mushroom-hunters.

Birke, Lynda, and Jo Hockenhull, eds. 2012. *Crossing Boundaries: Investigating Human-Animal Relationships*. Leiden, Netherlands: Brill.

Boer, Wietse de, Larissa B. Folman, Richard C. Summerbell, and Lynne Boddy. 2005. "Living in a Fungal World: Impact of Fungi on Soil Bacterial Niche Development." *FEMS Microbiology Reviews* 29, no. 4:795–811.

Bonfante, Paola. 2018. "The Future Has Roots in the Past: The Ideas and Scientists That Shaped Mycorrhizal Research." *New Phytologist* 220, no. 4:982–95.

Boyd, David R. 2012. *The Right to a Healthy Environment: Revitalizing Canada's Constitution*. Vancouver: University of British Columbia Press.

Bradley, David. 2001. "Language Policy for the Yi." In *Perspectives on the Yi of Southwest China*, edited by Stevan Harrell, 195–213. Berkeley: University of California Press.

Brandt, Jodi S., Tobias Kuemmerle, Haomin Li, Guopeng Ren, Jianguo Zhu, and Volker C. Radeloff. 2012. "Using Landsat Imagery to Map Forest Change in Southwest China in Response to the National Logging Ban and Ecotourism Development." *Remote Sensing of Environment* 121 (June): 358–69.

Brentari, Carlo. 2009. "Konrad Lorenz's Epistemological Criticism towards Jakob von Uexküll." *Σημειωτκή—Sign Systems Studies* 37, nos. 3–4:637–62.

Broucek, Jan. 2014. "Production of Methane Emissions from Ruminant Husbandry: A Review." *Journal of Environmental Protection* 5:1482–93.

Bunyard, Britt A. 2013. "Matsis and Wannabees: A Primer on Pine Mushrooms." *FUNGI* 6, no. 4:31–33.

Butler, Judith. 1988. "Performative Acts and Gender Constitution: An Essay in Phenomenology and Feminist Theory." *Theatre Journal* 40, no. 4:519–31.

Cadena, Marisol de la, and Mario Blaser, eds. 2018. *A World of Many Worlds*. Durham, NC: Duke University Press.

Caillois, Roger. (1970) 1985. *The Writing of Stones*. Charlottesville: University of Virginia Press.

Chamovitz, Daniel. 2012. *What a Plant Knows: A Field Guide to the Senses*. New York: Scientific American / Farrar, Straus and Giroux.

Chen, F. H., G. H. Dong, D. J. Zhang, X. Y. Liu, X. Jia, C. B. An, M. M. Ma, Y. W. Xie, L. Barton, X. Y. Ren, Z. J. Zhao, X. H. Wu, M. K. Jones. 2015. "Agriculture Facilitated Permanent Human Occupation of the Tibetan Plateau after 3600 B.P." *Science* 347, no. 6219:248–50.

Chen, Laurie. 2017. "Chinese Villager Finds the 'King of Mushrooms.'" *South China Morning Post*, October 23, 2017. https://www.scmp.com/news/china/society/article/2116557/chinese-villagers-find-king-mushrooms.

Cheney, D. L. and R. M. Seyfarth. 1990. *How Monkeys See the World: Inside the Mind of Another Species*. Chicago: University of Chicago Press.

Cheplick, Gregory Paul. 2015. *Approaches to Plant Evolutionary Ecology*. New York: Oxford University Press.

Cheung, Tobias. 2010. "What Is an 'Organism'"? On the Occurrence of a New Term and Its Conceptual Transformations 1680–1850." *History and Philosophy of the Life Sciences* 32, nos. 2–3:155–94.

Choy, Timothy. 2018. "Tending to Suspension: Abstraction and Apparatuses of Atmospheric Attunement in Matsutake Worlds." *Social Analysis* 62, no. 4:54–77.

Choy, Timothy K., Lieba Faier, Michael J. Hathaway, Miyako Inoue, Shiho Satsuka, and Anna Tsing. 2009. "A New Form of Collaboration in Cultural Anthropology: Matsutake Worlds." *American Ethnologist* 36, no. 2:380–403.

Chrulew, Matthew. 2020. "Reconstructing the Worlds of Wildlife: Uexküll, Hediger, and Beyond." *Biosemiotics* 13:137–49.

Classen, Constance. 1993. *Worlds of Sense: Exploring the Senses in History and across Cultures*. New York: Routledge.

Cleal, Christopher J, and Barry A. Thomas. 2005. "Palaeozoic Tropical Rainforests and Their Effect on Global Climates: Is the Past the Key to the Present?" *Geobiology* 3, no. 1:13–31.

Cline, Krista Marie Clark. 2010. "Psychological Effects of Dog Ownership: Role Strain, Role Enhancement, and Depression." *Journal of Social Psychology* 150, no. 2:117–31.

Corrales, Adriana, Scott A. Mangan, Benjamin L. Turner, and James W. Dalling. 2016. "An Ectomycorrhizal Nitrogen Economy Facilitates Monodominance in a Neotropical Forest." *Ecology Letters* 19, no. 4:383–92.

Crist, Eileen. *Images of Animals*. (1999) 2010. Philadelphia: Temple University Press.

Cromack K., P. Sollins, W. C. Graustein, K. Speidel, A. W. Todd, G. Spycher, C. Y. Li, and R. L. Todd. 1979. "Calcium Oxalate Accumulation and Soil Weathering in Mats of *Hysterangium crassum*." *Soil Biology and Biochemistry* 11:463–68.

Darwin, Charles. 1892. *The Formation of Vegetable Mould through the Action of Worms: With Observations on Their Habits.* Vol. 37. New York: Appleton.

Davis, Heather, and Zoe Todd. "On the Importance of a Date, or Decolonizing the Anthropocene." 2017. *ACME: An International E-Journal for Critical Geographies* 16, no. 4:761–80. https://acme-journal.org/index.php/acme/article/view/1539.

Dawkins, Richard. 1976. *The Selfish Gene.* Oxford: Oxford University Press.

———. 2004. "Extended Phenotype–but Not Too Extended: A Reply to Laland, Turner and Jablonka." *Biology and Philosophy* 19, no. 3:377–96.

Deleuze, Gilles. 1987. *A Thousand Plateaus: Capitalism and Schizophrenia.* Minneapolis: University of Minnesota Press.

De Queiroz, Kevin. 2005. "Ernst Mayr and the Modern Concept of Species." *Proceedings of the National Academy of Sciences* 102, no. 1:6600–6607.

De Waal, Frans. 1999. "Anthropomorphism and Anthropodenial: Consistency in Our Thinking about Humans and Other Animals." *Philosophical Topics* 27, no. 1:255–80.

———. 2008. *The Ape and the Sushi Master: Cultural Reflections of a Primatologist.* New York: Basic Books.

Despret, Vinciane. 2016. *What Would Animals Say if We Asked the Right Questions?* Translated by Brett Buchanan. Minneapolis: University of Minnesota Press.

Doty, Richard L. 2014. "Human Pheromones: Do They Exist." In *Neurobiology of Chemical Communication*, edited by Carla Mucignat-Caretta, 535–60. Boca Raton, FL: CRC.

Dunn, Rob. 2011. "Killer Fungi Made Us Hot-Blooded." *New Scientist*, November 30. https://www.newscientist.com/article/mg21228411-700-killer-fungi-made-us-hot-blooded/.

———. 2012. "Five Kinds of Fungus Discovered to Be Capable of Farming Animals!" *Scientific American*, February 20. https://blogs.scientificamerican.com/guest-blog/five-kinds-of-fungus-discovered-to-be-capable-of-farming-animals/.

———. 2018. *Never Home Alone: From Microbes to Millipedes, Camel Crickets, and Honeybees, the Natural History of Where We Live.* New York: Basic Books.

Dutton, Diane. 2012. "Being-with-Animals: Modes of Embodiment in Human-Animal Encounters." In *Crossing Boundaries: Investigating Human-Animal Relationships*, edited by Lynda Birke and Jo Hockenhull, 89–111. Leiden, Netherlands: Brill.

Elliott, T. F., D. S. Bower, and K. Vernes. 2019. "Reptilian Mycophagy: A Global Review of Mutually Beneficial Associations between Reptiles and Macrofungi." *Mycosphere* 10, no. 1:776–97.

Escobar, Arturo. 2018. *Designs for the Pluriverse: Radical Interdependence, Autonomy, and the Making of Worlds.* Durham, NC: Duke University Press.

Faier, Lieba. 2018. "Introduction: Elusive Matsutake." *Social Analysis* 62, no. 4:1–16.

Faier, Lieba, and Michael Hathaway, eds. 2018. "Matsutake Worlds" Special issue, articles by the Matsutake Worlds Research Group (Tim Choy, Lieba Faier, Elaine Gan, Michael Hathaway, Shiho Satsuka, and Anna Tsing). Afterword by Eduardo Kohn. *Social Analysis* 62, no. 4.

Faier, Lieba and Michael Hathaway, eds. *Matsutake Worlds*, Berghan, 2021

Faier, Lieba, for the Matsutake Worlds Research Group. 2011. "Fungi, Trees, People, Nematodes, Beetles, and Weather: Ecologies of Vulnerability and Ecologies of Negotiation in Matsutake Commodity Exchange." *Environment and Planning A* 43, no. 5:1079–97.

Faier, Lieba, and Lisa Rofel. 2014. "Ethnographies of Encounter." *Annual Review of Anthropology* 43:363–77.

Fang, R., P. Kirk, J. C. Wei, Y. Li, L. Cai, L. Fan, T. Z. Wei, R. L. Zhao, K. Wang, and Z. L. Yang. 2018. "Country Focus: China." In *State of the World's Fungi*, edited by K. J. Willis, 48–55. Richmond, UK: Royal Botanic Gardens, Kew.

Favareau, Donald, ed. 2009. *Essential Readings in Biosemiotics*. New York: Springer.

Faye. 2016 "Domestic Market Makes Up for China's Weak Matsutake Exports." *Produce Report*, October 9. https://www.producereport.com/article/domestic-market-makes-chinas-weak -matsutake-exports.

Felton, Gary W., and James H. Tumlinson. 2008. "Plant-Insect Dialogs: Complex Interactions at the Plant-Insect Interface." *Current Opinion in Plant Biology* 11, no. 4:457–63.

Finch, Caleb E. 1994. *Longevity, Senescence, and the Genome*. Chicago: University of Chicago Press.

Finlay, Roger D. 2008. "Ecological Aspects of Mycorrhizal Symbiosis: With Special Emphasis on the Functional Diversity of Interactions Involving the Extraradical Mycelium." *Journal of Experimental Botany* 59, no. 5:1115–26.

Fiskesjö, Magnus. 1999. "On the 'Raw' and the 'Cooked' Barbarians of Imperial China." *Inner Asia* 1, no. 2:139–68.

Francis, Claire. 2020. "Sensory Trust.5, 9, 21, 53 . . . How Many Senses?" *Sensory Trust*, October 15. https://www.sensorytrust.org.uk/information/articles/senses.html.

Frank, Albert Bernhard. 1885. "Über die auf Wurzelsymbiose beruhende Ernährung gewisser Bäume durch unterirdische Pilze" [On the nourishing, via root symbiosis, of certain trees by underground fungi]. *Berichte der Deutschen Botanischen Gesellschaft* 3:128–45.

Frank, Mark E. 2018. "Hacking the Yak." *East Asian Science, Technology, and Medicine*, no. 48:17–48.

Frazer, Jennifer. 2011. "A New Weapon in the War on Frog Chytrids: Water Fleas." *Scientific American*, September 2. https://blogs.scientificamerican.com/artful-amoeba/a-new -weapon-in-the-war-on-frog-chytrids-water-fleas/.

———. 2015. "The World's Largest Mining Operation Is Run by Fungi." *Scientific American*, November 5. https://blogs.scientificamerican.com/artful-amoeba/the-world-s-largest -mining-operation-is-run-by-fungi/.

Gadd, Geoffrey M. 2007. "Geomycology: Biogeochemical Transformations of Rocks, Minerals, Metals and Radionuclides by Fungi, Bioweathering and Bioremediation." *Mycological Research* 111, no. 1:3–49.

Gagliano, Monica. 2017. "The Mind of Plants: Thinking the Unthinkable." *Communicative and Integrative Biology* 10, no. 2:e1288333–e1288333-4.

Gaycken, Oliver. 2012. "The Secret Life of Plants: Visualizing Vegetative Movement, 1880–1903." *Early Popular Visual Culture* 10, no. 1:51–69.

George, Justin, Nina E. Jenkins, Simon Blanford, Matthew B. Thomas, and Thomas C. Baker. 2013. "Malaria Mosquitoes Attracted by Fatal Fungus." *PLOS One* 8, no. 5:e62632.

Ghannoum, Mahmoud A., and Pranab K. Mukherjee. 2013. "The Human Mycobiome and Its Impact on Health and Disease." *Current Fungal Infection Reports* 7, no. 4:345–50.

Giffney, Noreen, and Myra J. Hird, eds. 2008. *Queering the Non/Human*. Burlington, VT: Ashgate.

Gillen, Paul. 2016. "Notes on Mineral Evolution: Life, Sentience, and the Anthropocene." *Environmental Humanities* 8, no. 2:215–34.

Glass, N. Louise, Carolyn Rasmussen, M. Gabriela Roca, and Nick D. Read. 2004. "Hyphal Homing, Fusion and Mycelial Interconnectedness." *Trends in Microbiology* 12, no. 3:135–41.

Glauben, Thomas, Thomas Herzfeld, Scott Rozelle, and Xiaobing Wang. 2012. "Persistent Poverty in Rural China: Where, Why, and How to Escape?" *World Development* 40, no. 4:784–95.

Godfrey-Smith, Peter. 2016. *Other Minds: The Octopus, the Sea, and the Deep Origins of Consciousness*. New York: Farrar, Straus and Giroux.

Golan, Jacob J., and Anne Pringle. 2017. "Long-Distance Dispersal of Fungi." *Microbiology Spectrum* 5, no. 4. https://doi.org/10.1128/microbiolspec.FUNK-0047-2016.

Goldstein, Melvyn C. 1997. *The Snow Lion and the Dragon: China, Tibet, and the Dalai Lama*. Berkeley: University of California Press.

Gong, Ming-Qin, and Feng-Zhen Wang. 2004. "The Countermeasures of China to Present Market Status of *Tricholoma matsutake*" (in Chinese). *Territory and Natural Resources Study* 2:88–89.

Greenberg, Dan A., and Wendy J. Palen. 2019. "A Deadly Amphibian Disease Goes Global." *Science* 363, no. 6434:1386–88.

Griffiths, David. 2015. "Queer Theory for Lichens." *UnderCurrents: Journal of Critical Environmental Studies* 19:36–45.

Grime, J. P., J.M.L. Mackey, S. H. Hillier, and D. J. Read. 1987. "Floristic Diversity in a Model System Using Experimental Microcosms." *Nature* 328, no. 6129:420–22.

Grosz, Elizabeth. 2011. *Becoming Undone: Darwinian Reflections on Life, Politics, and Art*. Durham, NC: Duke University Press.

Gruen, Lori. 2011. *Ethics and Animals: An Introduction*. Cambridge: Cambridge University Press.

———. 2015. *Entangled Empathy: An Alternative Ethic for Our Relationships with Animals*. New York: Lantern Books.

Guo, H. J., and C. L. Long. 1998. *Yunnan's Biodiversity*. Kunming, China: Yunnan Science and Technology Press (in Chinese).

Guo, Yanlong, Xin Li, Zefang Zhao, Haiyan Wei, Bei Gao, and Wei Gu. 2017. "Prediction of the Potential Geographic Distribution of the Ectomycorrhizal Mushroom *Tricholoma matsutake* under Multiple Climate Change Scenarios." *Scientific Reports* 7, no. 46221:1–11.

Hallé, Francis. (1999) 2011. *In Praise of Plants*. Translated by David Lee. Portland, OR: Timber.

Haraway, Donna J. 1997. *Modest_Witness@Second_Millennium.FemaleMan_Meets_OncoMouse*. New York: Routledge.

———. 2003. *The Companion Species Manifesto: Dogs, People, and Significant Otherness*. Vol. 1. Chicago: Prickly Paradigm.

———. (2007) 2013. *When Species Meet*. Minneapolis: University of Minnesota Press.

———. 2016. *Staying with the Trouble: Making Kin in the Chthulucene*. Durham, NC: Duke University Press.

Harrell, Stevan, ed. 2001. *Perspectives on the Yi of Southwest China*. Berkeley: University of California Press.

Hassett, Maribeth O., Mark W. F. Fischer, and Nicholas P. Money. 2015. "Mushrooms as Rainmakers: How Spores Act as Nuclei for Raindrops." *PLOS One* 10, no. 10:1–10.

Hathaway, Michael J. 2013. *Environmental Winds: Making the Global in Southwest China.* Berkeley: University of California Press.

———. 2017. "Discovering China's Tropical Rainforests: Shifting Approaches to People and Nature in the Late Twentieth Century." In *The Nature State: Rethinking the History of Conservation,* edited by Wilko Graf von Hardenberg, Matthew Kelly, Claudia Leal, and Emily Wakild, 176–93. London: Routledge.

Hay, William Delisle. 1887. *An Elementary Text-Book of British Fungi.* London: S. Sonnenschein, Lowrey.

Hayes, Jack Patrick. 2013. *A Change in Worlds on the Sino-Tibetan Borderlands: Politics, Economies, and Environments in Northern Sichuan.* Washington, DC: Lexington Books.

Hayward, Eva. 2010. "Fingeryeyes: Impressions of Cup Corals." *Cultural Anthropology* 25, no. 4:577–99.

Hazen, Robert M., Dominic Papineau, Wouter Bleeker, Robert T. Downs, John M. Ferry, Timothy J. McCoy, Dimitri A. Sverjensky, and Hexiong Yang. 2008. "Mineral Evolution." *American Mineralogist* 93, nos. 11–12:1693–720.

He, Jun. 2010. "Globalised Forest-Products: Commodification of the Matsutake Mushroom in Tibetan villages, Yunnan, Southwest China." *International Forestry Review* 12 no. 1:27–37.

Healy, Kevin, Luke McNally, Graeme D. Ruxton, Natalie Cooper, and Andrew L. Jackson. 2013. "Metabolic Rate and Body Size Are Linked with Perception of Temporal Information." *Animal Behaviour* 86, no. 4:685–96.

Heckman, Daniel S., David M. Geiser, Brooke R. Eidell, Rebecca L. Stauffer, Natalie L. Kardos, and S. Blair Hedges. 2001. "Molecular Evidence for the Early Colonization of Land by Fungi and Plants." *Science* 293, no. 5532:1129–33.

Hehemann, Jan-Hendrik, Gaëlle Correc, Tristan Barbeyron, William Helbert, Mirjam Czjzek, and Gurvan Michel. 2010. "Transfer of Carbohydrate-Active Enzymes from Marine Bacteria to Japanese Gut Microbiota." *Nature* 464, no. 7290:908–12.

Heidegger, Martin. 1995. *The Fundamental Concepts of Metaphysics: World, Finitude, Solitude.* Translated by William McNeill and Nicholas Walker. Bloomington: Indiana University Press.

Henchman, Anna A. 2017. "Charles Darwin's Final Book on Earthworms, 1881." *BRANCH: Britain, Representation, and Nineteenth-Century History.* http://www.branchcollective.org/?ps _articles=anna-henchman-charles-darwins-final-book-on-earthworms-1881.

Hird, Myra J. 2000. "Gender's Nature: Intersexuality, Transsexualism and the 'Sex'/'Gender' Binary." *Feminist Theory* 1, no. 3:347–64.

———. 2004. *Sex, Gender, and Science.* New York: Palgrave Macmillan.

Hiromoto, K. 1963. "Life-Relation between *Pinus densiflora* and *Armillaria matsutake* II: Mycorrhiza of *Pinus densiflora* with *Armillaria matsutake.*" *Botanical Magazine* 76:292–98 (in Japanese).

Hochberg, Michael E., Guillaume Bertault, Karine Poitrineau, and Arne Janssen. 2003. "Olfactory Orientation of the Truffle Beetle, *Leiodes cinnamomea.*" *Entomologia Experimentalis et Applicata* 109, no. 2:147–53.

Horowitz, Alexandra. 2010. *Inside of a Dog: What Dogs See, Smell, and Know.* New York: Simon and Schuster.

———. 2016. *Being a Dog: Following the Dog into a World of Smell*. New York: Simon and Schuster.

Hou, Yiling, Xiang Ding, Wanru Hou, Jie Zhong, Hongqing Zhu, Binxiang Ma, Ting Xu, and Junhua Li. 2013. "Anti-microorganism, Anti-tumor, and Immune Activities of a Novel Polysaccharide Isolated from *Tricholoma matsutake*." *Pharmacognosy Magazine* 9, no. 35:244–49.

Hoysted, Grace A., Jill Kowal, Alison Jacob, William R. Rimington, Jeffrey G. Duckett, Silvia Pressel, Suzanne Orchard, Megan H. Ryan, Katie J. Field, and Martin I. Bidartondo. 2018. "A Mycorrhizal Revolution." *Current Opinion in Plant Biology* 44:1–6.

Hribal, Jason C. 2000. "Animals, Agency, and Class: Writing the History of Animals from Below." *Human Ecology Review* 14, no. 1:101–12.

Hughes, Sylvia. 1990. "Antelope Activate the Acacia's Alarm System." *New Scientist*. September 28. https://www.newscientist.com/article/mg12717361-200-antelope-activate-the-acacias-alarm-system/.

Hustak, Carla, and Natasha Myers. 2012. "Involutionary Momentum: Affective Ecologies and the Sciences of Plant/Insect Encounters." *Differences* 23, no. 3:74–118.

Huttner-Koros, A. 2015. "The Hidden Bias of Science's Universal Language." *Atlantic*, August 21. https://www.theatlantic.com/science/archive/2015/08/english-universal-language-science-research/400919/.

Huxley, Julian. 1912. *The Individual in the Animal Kingdom*. Cambridge: Cambridge University Press.

Hyde, Kevin D., E. B. Gareth Jones, Eduardo Leaño, Stephen B. Pointing, Asha D. Poonyth, and Lilian L. P. Vrijmoed. 1998. "Role of Fungi in Marine Ecosystems." *Biodiversity and Conservation* 7, no. 9:1147–61.

Hyde, Sandra Teresa. 2007. *Eating Spring Rice: The Cultural Politics of AIDS in Southwest China*. Berkeley: University of California Press.

Ingold, Tim. 2000. *The Perception of the Environment: Essays on Livelihood, Dwelling and Skill*. New York: Psychology.

———. 2011. *Being Alive: Essays on Movement, Knowledge and Description*. New York: Taylor and Francis.

———. 2015. *The Life of Lines*. New York: Routledge.

Israel, John. 1999. *Lianda: A Chinese University in War and Revolution*. Palo Alto, CA: Stanford University Press.

Ittner, Lukas D., Marion Junghans, and Inge Werner. 2018. "Aquatic Fungi: A Disregarded Trophic Level in Ecological Risk Assessment of Organic Fungicides." *Frontiers in Environmental Science* 6:105.

Ivarsson, Magnus, Stefan Bengtson, and Anna Neubeck. 2016. "The Igneous Oceanic Crust—Earth's Largest Fungal Habitat?" *Fungal Ecology* 20:249–55.

Jakobsen, I., L. K. Abbott, and A. D. Robson. 1992. "External Hyphae of Vesicular-Arbuscular Mycorrhizal Fungi Associated with *Trifolium subterraneum* L. 1. Spread of Hyphae and Phosphorus Inflow into Roots." *New Phytologist* 120, no. 3:371–80.

Jennersten, Ola. 1988. "Insect Dispersal of Fungal Disease: Effects of *Ustilago* Infection on Pollinator Attraction in *Viscaria vulgaris*." *Oikos* 51, no. 2:163–70.

Jennings, David Harry. 1995. *The Physiology of Fungal Nutrition*. Cambridge: Cambridge University Press.

Jenssen, Thomas A. 1967. "Food Habits of the Green Frog, *Rana clamitans*, before and during Metamorphosis." *Copeia*, no. 1:214–18.

Johnson, Melissa. 2018. *Becoming Creole: Nature and Race in Belize*. New Brunswick, NJ: Rutgers University Press.

Jongmans, A. G., N. Van Breemen, U. Lundström, P.A.W. Van Hees, R. D. Finlay, M. Srinivasan, T. Unestam, R. Giesler, P. A. Melkerud, and M. Olsson. 1997. "Rock-Eating Fungi." *Nature* 389, no. 6652:682–83.

Joshi, S., L. Shrestha, N. Bisht, N. Wu, M. Ismail, T. Dorii, G. Dangol, and R. Long. 2020. "Ethnic and Cultural Diversity amongst Yak Herding Communities in the Asian Highlands." *Sustainability* 12, no. 3:957.

Jussieu, M. M., and Ad. Brongniart. 1851. "II.—Report on MM L. R. and C. Tulasne's 'Memoir on the History of the Hypogæous Fungi.'" *Annals and Magazine of Natural History* 8, no. 43:19–25.

Kant, Immanuel. 1798. *Kant: Anthropology from a Pragmatic Point of View*. Cambridge: Cambridge University Press.

Keller, Evelyn Fox. 2008. "Organisms, Machines, and Thunderstorms: A History of Self-Organization, Part One." *Historical Studies in the Natural Sciences* 38, no. 1:45–75.

Kemmerer, Lisa. 2006. "Verbal Activism: 'Anymal.'" *Society and Animals* 14, no. 1:9–14.

Kendrick, Bryce. 2011. *Fungi: Ecological Importance and Impact on Humans*. Chichester, UK: John Wiley and Sons.

Kenrick, Paul, and Christine Strullu-Derrien. 2014. "The Origin and Early Evolution of Roots." *Plant Physiology* 166, no. 2:570–80.

Khan, Sehroon, Sadia Nadir, Zia Ullah Shah, Aamer Ali Shah, Samantha C. Karunarathna, Jianchu Xu, Afsar Khan, Shahzad Munir, and Fariha Hasan. 2017. "Biodegradation of Polyester Polyurethane by *Aspergillus tubingensis*." *Environmental Pollution* 225:469–80.

Kim, N. 2018. "Fuel for the Smelters: Copper Mining and Deforestation in Northeastern Yunnan during the High Qing, 1700 to 1850." In *Southwest China in a Regional and Global Perspective (c. 1600–1911)*, edited by Ulrich Theobald and Jin Cao, 87–123. Leiden: Brill.

Kimmerer, Robin Wall. 2003. *Gathering Moss: A Natural and Cultural History of Mosses*. Corvalis: Oregon State University Press.

———. 2013. *Braiding Sweetgrass: Indigenous Wisdom, Scientific Knowledge and the Teachings of Plants*. Minneapolis: Milkweed Editions.

King, John. 1997. *Reaching for the Sun: How Plants Work*. Cambridge: Cambridge University Press.

Kinugawa, K., and T. Goto. 1978. "Preliminary Survey on the Matsutake (*Armillaria ponderosa*) of North America." *Transactions of the Mycological Society of Japan* 19:91–101.

Kirksey, S. Eben, and Stefan Helmreich. 2010. "The Emergence of Multispecies Ethnography." *Cultural Anthropology* 25, no. 4:545–76.

Kish, Daniel. 2009. "Human Echolocation: How to 'See' Like a Bat." *New Scientist* 202, no. 2703:31–33.

Kohn, Eduard. 2013. *How Forests Think: Toward an Anthropology beyond the Human*. Berkeley: University of California Press.

Kolas, Ashild. 2007. *Tourism and Tibetan Culture in Transition*. New York: Routledge.

Koivusaari, Pirjo, Mysore V. Tejesvi, Mikko Tolkkinen, Annamari Markkola, Heikki Mykrä, and Anna Maria Pirttilä. 2019. "Fungi originating from tree leaves contribute to fungal diversity of litter in streams." *Frontiers in Microbiology* 10:651.

Krebs, Charles, and Briana Elwood. 2008. *The Ecological World View*. Berkeley: University of California Press.

Krings, Michael, Thomas N. Taylor, Hagen Hass, Hans Kerp, Nora Dotzler, Elizabeth J. Hermsen. 2007. "Fungal Endophytes in a 400-Million-Yr-Old Land Plant: Infection Pathways, Spatial Distribution, and Host Responses." *New Phytologist* 174, no. 3:648–57.

Krützen, Michael, Carel van Schaik, and Andrew Whiten. 2007. "The Animal Cultures Debate: Response to Laland and Janik." *Trends in Ecology and Evolution* 22, no. 1:6.

Laland, Kevin N. 2004. "Extending the Extended Phenotype." *Biology and Philosophy* 19, no. 3:313–25.

Laland, Kevin N., and Bennett G. Galef. 2009. *The Question of Animal Culture*. Cambridge, MA: Harvard University Press.

Laland, Kevin N., and Vincent M. Janik. 2006. "The Animal Cultures Debate." *Trends in Ecology and Evolution* 21, no. 10:542–47.

Landecker, Hannah. 2016. "Antibiotic Resistance and the Biology of History." *Body and Society* 22, no. 4:19–52.

Latour, Bruno. 1984. *Les Microbes Guerre et Paix: Suivi de, Irréductions*. Paris: Editions Métailié.

———. 1992. "Where Are the Missing Masses? The Sociology of a Few Mundane Artifacts." In *Shaping Technology / Building Society: Studies in Sociotechnical Change*, edited by Wiebe E. Bijker and John Law, 225–58. Cambridge, MA: MIT Press.

Laurent, Yannick. 2015. "The Tibetans in the Making: Barley Cultivation and Cultural Representations." *Revue d'Etudes Tibétaines*, no. 33:73–108.

Leake, Jonathan, David Johnson, Damian Donnelly, Gemma Muckle, Lynne Boddy, and David Read. 2004. "Networks of Power and Influence: The Role of Mycorrhizal Mycelium in Controlling Plant Communities and Agroecosystem Functioning." *Canadian Journal of Botany* 82, no. 8:1016–45.

Lefevre, Charles K. 2003. "Host Associations of *Tricholoma magnivelare*, the American Matsutake." PhD diss., Department of Forest Science, Oregon State University.

Lévi-Strauss, Claude. 1963. *Le Totémisme aujourd'hui* [Totemism]. Boston: Beacon.

Li, Bo. 2002. "'The Lost Horizon': In Search of Community-Based Natural Resource Management in Nature Reserves of Northwest Yunnan, China." Master's thesis, Department of Natural Resources, Cornell University.

Li, Jianglin. 2016. *Tibet in Agony: Lhasa 1959*. Cambridge, MA: Harvard University Press.

Li, Kun-Sheng. 2004. "1949年以来云南人类起源与史前考古的主要成就 [A summary of the main achievements of the Mankind Origin Study and the prehistoric archaeology in Yunnan]." *云南社会科学* 2:103–13 and 136.

Li, Shu-Hong, Hong-Mei Chai, Kai-Mei Su, Ming-Hui Zhong, and Yong-Chang Zhao. 2010. "Resources Investigation and Sustainable Suggestions on the Wild Mushrooms in Jianchuan." *Edible Fungi of China* 5:7–11.

Li, Zibo, Lianwen Liu, Jun Chen, and H. Henry Teng. 2016. "Cellular Dissolution at Hypha- and Spore-Mineral Interfaces Revealing Unrecognized Mechanisms and Scales of Fungal Weathering." *Geology* 44, no. 4:319–22.

Lien, Marianne Elisabeth. 2015. *Becoming Salmon: Aquaculture and the Domestication of a Fish.* Berkeley: University of California Press.

Lien, Marianne Elisabeth, and Gisli Pálsson. 2019. "Ethnography beyond the Human: The 'Other-than-Human' in Ethnographic Work." *Ethnos* 86, no. 3:1–20.

Liu, Xinyi, Diane L. Lister, Zhijun Zhao, Cameron A. Petrie, Xiongsheng Zeng, Penelope J. Jones, Richard A. Staff, Anil K. Pokharia, Jennifer Bates, and Ravindra N. Singh. 2017. "Journey to the East: Diverse Routes and Variable Flowering Times for Wheat and Barley En Route to Prehistoric China." *PLOS One* 12, no. 11:e0209518.

Lorimer, Jamie. 2007. "Nonhuman Charisma." *Environment and Planning D: Society and Space* 25, no. 5:911–32.

———. 2015. *Wildlife in the Anthropocene: Conservation after Nature.* Minneapolis: University of Minnesota Press.

Loron, Corentin C., Camille François, Robert H. Rainbird, Elizabeth C. Turner, Stephan Borensztajn, and Emmanuelle J. Javaux. 2019. "Early Fungi from the Proterozoic Era in Arctic Canada." *Nature* 570, no. 7760:232–35.

Luginbuehl, Leonie H., and Giles E. D. Oldroyd. 2017. "Understanding the Arbuscule at the Heart of Endomycorrhizal Symbioses in Plants." *Current Biology* 27, no. 17:R952–63.

Macfarlane, Robert. 2016. "The Secrets of the Wood Wide Web." *New Yorker*, August 7. https://www.newyorker.com/tech/annals-of-technology/the-secrets-of-the-wood-wide-web.

MacMurray, Jessica. 2003. "Matsutake Gari: Hunting for Mushrooms in a New West." *Gastronomica* 3, no. 4:86–89.

Magnus, Riin. 2014. "Training Guide Dogs of the Blind with the 'Phantom Man' Method: Historic Background and Semiotic Footing." *Semiotica* 2014, no. 198:181–204.

Mahmood, Saba. 2011. *The Politics of Piety: The Islamic Revival and the Feminist Subject.* Princeton, NJ: Princeton University Press.

Makley, Charlene E. 1997. "The Meaning of Liberation: Representations of Tibetan Women." *Tibet Journal* 22, no. 2:4–29.

———. 2007. *The Violence of Liberation: Gender and Tibetan Buddhist Revival in Post-Mao China.* Berkeley: University of California Press.

Maran, Timo, Dario Martinelli, and Aleksei Turovski, eds. 2012. *Readings in Zoosemiotics.* Berlin: Walter de Gruyter.

Margulis, Lynn. 1998. *Symbiotic Planet: A New Look at Evolution.* New York: Basic Books.

———. 1990. "Words as Battle Cries: Symbiogenesis and the New Field of Endocytobiology." *Bioscience* 40, no. 9:673–77.

Margulis, Lynn, and Dorion Sagan. 1997. *Microcosmos: Four Billion Years of Microbial Evolution.* Berkeley: University of California Press.

Martin, Emily. 1991. "The Egg and the Sperm: How Science Has Constructed a Romance Based on Stereotypical Male-Female Roles." *Signs: Journal of Women in Culture and Society* 16, no. 3:485–501.

———. 2001. *The Woman in the Body: A Cultural Analysis of Reproduction.* Boston: Beacon.

Maser, Chris, Andrew W. Claridge, and James M. Trappe. 2008. *Trees, Truffles, and Beasts: How Forests Function.* New Brunswick, NJ: Rutgers University Press.

Masson, Jeffrey M., and Susan McCarthy. 1996. *When Elephants Weep: The Emotional Lives of Animals.* New York: Random House.

Mather, Jennifer A., Roland C. Anderson, and James B. Wood. 2013. *Octopus: The Ocean's Intelligent Invertebrate.* Portland, OR: Timber.

Matheson, Thomas. 2001. *Invertebrate Nervous Systems.* Chichester, UK: John Wiley and Sons.

Matsutake Worlds Research Group (Tim Choy, Lieba Faier, Michael Hathaway, Miyako Inoue, Shiho Satsuka and Anna Tsing). 2009a. "A New Form of Collaboration in Cultural Anthropology: Matsutake Worlds." *American Ethnologist* 36, no. 2:380–403.

———. 2009b. "Strong Collaboration as Method for Multi-sited Ethnography: On Mycorrhizal Relations" In *Multi-sited Ethnography: Theory, Praxis, and Locality in Contemporary Social Research,* edited by Mark-Anthony Falzon, 197–214. New York: Routledge.

———. 2018. "Matsutake Worlds." *Social Analysis* 62, no. 4. https://www.berghahnjournals.com /view/journals/social-analysis/62/4/social-analysis.62.issue-4.xml.

McCarthy, Susan. 2011. *Communist Multiculturalism: Ethnic Revival in Southwest China.* Seattle: University of Washington Press.

McCormick, Andrea Clavijo, Sybille B. Unsicker, and Jonathan Gershenzon. 2012. "The Specificity of Herbivore-Induced Plant Volatiles in Attracting Herbivore Enemies." *Trends in Plant Science* 17, no. 5:303–10.

McCoy, Peter. 2016. *Radical Mycology: A Treatise on Seeing and Working with Fungi.* Portland, OR: Chthaeus.

McFall-Ngai, Margaret, Michael G. Hadfield, Thomas C. G. Bosch, Hannah V. Carey, Tomislav Domazet-Lošo, Angela E. Douglas, Nicole Dubilier, Gerard Eberl, Tadashi Fukami, and Scott F. Gilbert. 2013. "Animals in a Bacterial World, a New Imperative for the Life Sciences." *Proceedings of the National Academy of Sciences* 110, no. 9:3229–36.

McGranahan, Carole. 2005. "Truth, Fear, and Lies: Exile Politics and Arrested Histories of the Tibetan Resistance." *Cultural Anthropology* 20, no. 4:570–600.

———. 2010. *Arrested Histories: Tibet, the CIA, and Memories of a Forgotten War.* Durham, NC: Duke University Press.

———. 2019. "Chinese Settler Colonialism: Empire and Life in the Tibetan Borderlands." In *Frontier Tibet,* edited by Stéphane Gros, 518–39. Amsterdam, Netherlands: Amsterdam University Press.

Melin, Elias. 1925. *Untersuchungen über die bedeutung der baummykorrhiza, eine ökologisch-physiologische studie* [Studies on the importance of tree mycorrhiza, an ecological-physiological study]. Stuttgart, Germany: G. Fischer, Jena.

Menzies, Nicholas. 1994. *Forest and Land Management in Imperial China.* Basingstoke, UK: Macmillan.

Miller, D. 1999. "Nomads of Tibetan Rangelands in Western China." *Rangelands* 21, no. 1:16–19.

Miller, Timothy. n.d. "Botanical Fiction Database." *Timothy S. Miller, Ph.D.* Accessed March 22, 2020. http://www.thefishinprison.com/botanical-fiction-database.html.

Millman, Lawrence. 2019. *Fungipedia: A Brief Compendium of Mushroom Lore.* Princeton, NJ: Princeton University Press.

Molnar, Charles, and Jane Gair. 2015. *Concepts of Biology*. 1st Canadian ed. Vancouver: BCcampus, OpenEd. https://openlibrary-repo.ecampusontario.ca/jspui/bitstream/123456789/345/5/Concepts-of-Biology-1st-Canadian-Edition-1514999939.html.

Money, Nicholas P., as told to Flora Lichtman. 2011. "How Fungi Launch Spores Is No Longer a Mystery." *Popular Science*, June 6. https://www.popsci.com/science/article/2011-05/how-fungi-launch-spores-no-longer-mystery/.

———. 2016. "Women Mycologists." *OUP Blog*. https://blog.oup.com/2016/03/women-mycologists/.

Montgomery, Sy. 2015. *The Soul of an Octopus: A Surprising Exploration into the Wonder of Consciousness*. New York: Simon and Schuster.

Moran, Barbara. 2016. "Soil Fungi Could Affect Climate Change." *Brink*, November 15. https://www.bu.edu/articles/2016/soil-fungi/.

Morell, Virginia. 2016. "Woodpeckers Partner with Fungi to Build Homes." *ScienceMag*, March 22. https://www.sciencemag.org/news/2016/03/woodpeckers-partner-fungi-build-homes.

Morris, Brian. 1988. "The Folk Classification of Fungi." *Mycologist* 2, no. 1:8–10.

Mortensen, Dasa. 2016. "The History of Gyalthang under Chinese Rule: Memory, Identity, and Contested Control in a Tibetan Region of Northwest Yunnan." PhD diss., Department of History, University of North Carolina at Chapel Hill.

Mortensen, Eric D. 2016. "Prosperity, Identity, Intra-Tibetan Violence, and Harmony." In *Ethnic Conflict and Protest in Tibet and Xinjiang: Unrest in China's West*, edited by Ben Hillman and Gray Tuttle, 201–22. New York: Columbia University Press.

Mortimer-Sandilands, Catriona, and Bruce Erickson. 2010. *Queer Ecologies: Sex, Nature, Politics, Desire*. Bloomington: Indiana University Press.

Mueggler, Erik. 2001. *The Age of Wild Ghosts: Memory, Violence, and Place in Southwest China*. Berkeley: University of California Press.

Mueller, Ulrich G., and Nicole Gerardo. 2002. "Fungus-Farming Insects: Multiple Origins and Diverse Evolutionary Histories." *Proceedings of the National Academy of Sciences* 99, no. 24:15247–49.

Muldavin, Joshua. 2000. "The Geography of Japanese Development Aid to China, 1978–98." *Environment and Planning A* 32, no. 5:925–46.

Murdoch, David R., and Nigel P. French. 2020. "COVID-19: Another Infectious Disease Emerging at the Animal-Human Interface." *New Zealand Medical Journal* 133, no. 1510:12–15.

Myers, Natasha. 2015. "Conversations on Plant Sensing." *Nature Culture* 3:35–66.

Nagel, Thomas. 1974. "What Is It Like to Be a Bat?" *Philosophical Review* 83, no. 4:435–50.

Nakayama, Thomas K., and Charles E. Morris III. 2015. "Worldmaking and Everyday Interventions." *QED: A Journal in GLBTQ Worldmaking* 2, no. 1:v–viii.

Nelsen, Matthew P., William A. DiMichele, Shanan E. Peters, and C. Kevin Boyce. 2016. "Delayed Fungal Evolution Did Not Cause the Paleozoic Peak in Coal Production." *Proceedings of the National Academy of Sciences* 113, no. 9:2442–47.

Newman, E. I. 1988. "Mycorrhizal Links between Plants: Their Functioning and Ecological Significance." In *Advances in Ecological Research*, vol. 18, edited by M. Begon, Alastair Fitter, E. Ford, and A. Macfadyen, 243–70. New York: Academic.

Ogden, Laura A., Billy Hall, and Kimiko Tanita. 2013. "Animals, Plants, People, and Things: A Review of Multispecies Ethnography." *Environment and Society* 4, no. 1:5–24.

Omura, Kei'ichi, Grant Otsuki, Shiho Satsuka, and Atsuro Morita, eds. 2018. *The World Multiple: The Quotidian Politics of Knowing and Generating Entangled Worlds.* New York: Routledge.

Pálsson, Gísli. 1994. "Enskilment at Sea." *Man* 29, no. 4:901–27.

Patowary, Kaushik. 2017. "The Humongous Fungus." *Amusing Planet*, July 7. https://www .amusingplanet.com/2017/07/the-humongous-fungus.html.

Peters-Golden, Holly. 2012. *Culture Sketches: Case Studies in Anthropology.* New York: McGraw-Hill.

Philo, Chris, and Chris Wilbert. 2000. "Animal Spaces, Beastly Places: An Introduction." In *Animal Spaces, Beastly Places: New Geographies of Human-Animal Relations*, edited by Chris Philo and Chris Wilbert, 1–34. London: Routledge.

Pierson, J. L. 1958. *The Manyosu: Translated and Annotated.* Book 10. Leiden: E. J. Brill.

Plumwood, Val. 2002a. "Decolonizing Relationships with Nature." *PAN: Philosophy Activism Nature*, no. 2:7–30.

———. 2002b. *Feminism and the Mastery of Nature.* New York: Routledge.

Pollan, Michael. 2013. "The Intelligent Plant." *New Yorker*, December 23 and 30, 92–105.

Popova, Maria. 2015. "Diane Ackerman on the Secret Life of the Senses and the Measure of Our Aliveness." *Brain Pickings*, August 6. https://www.brainpickings.org/2015/08/06/diane -ackerman-a-natural-history-of-the-senses-2/.

Pouliot, Alison. 2016. "A Thousand Days in the Forest: An Ethnography of the Culture of Fungi." PhD diss., Department of Biodiversity Conservation, Australian National University.

Prestholdt, Jeremy. 2018. *Domesticating the World: East African Consumerism and the Genealogies of Globalization.* Berkeley: University of California Press.

Pringle, Anne. 2017. "Establishing New Worlds: The Lichens of Petersham." In *Arts of Living on a Damaged Planet: Ghosts and Monsters of the Anthropocene*, edited by Anna Lowenhaupt Tsing, Nils Bubandt, Elaine Gan, and Heather Anne Swanson, 157–68. Minneapolis: University of Minnesota Press.

Probyn-Rapsey, Fiona. 2018. "Anthropocentrism." In *Critical Terms in Animal Studies*, edited by Lori Gruen, 47–63. Chicago: University of Chicago Press.

Pyare, Sanjay, and William S. Longland. 2002. "Interrelationships among Northern Flying Squirrels, Truffles, and Microhabitat Structure in Sierra Nevada Old-Growth Habitat." *Canadian Journal of Forest Research* 32, no. 6:1016–24.

Qian, F., Q. Li, P. Wu, S. Yuan, R. Xing, H. Chen, and H. Zhang. 1991. "Lower Pleistocene, Yuanmou Formation." In *Quaternary Geology and Paleoanthropology of Yuanmou, Yunnan, China*, edited by F. Qian and G. Zhou, 17–50. Beijing, China: Science Press (in Chinese).

Raffles, Hugh. 2020. *The Book of Unconformities.* New York: Random House.

Raper, Cardy. 2013. *A Woman of Science: An Extraordinary Journey of Love, Discovery, and the Sex Life of Mushrooms.* Hobart, NY: Hatherleigh.

Rayner, Alan D. M. 1997. *Degrees of Freedom: Living in Dynamic Boundaries.* Vol. 23. London: Imperial College Press.

Redhead, Scott A. 1997. "The Pine Mushroom Industry in Canada and the United States: Why It Exists and Where It Is Going." In *Mycology in Sustainable Development: Expanding Concepts and Vanishing Borders*, edited by M. E. Palm and I. H. Chapela, 11–46. Blowing Rock, NC: Parkway.

Rhode, David, David B. Madsen, P. Jeffrey Brantingham, and Tsultrim Dargye. 2007. "Yaks, Yak Dung, and Prehistoric Human Habitation of the Tibetan Plateau." *Developments in Quaternary Sciences* 9:205–24.

Riskin, Jessica. 2016. *The Restless Clock: A History of the Centuries-Long Argument over What Makes Living Things Tick*. Chicago: University of Chicago Press.

Robinson, D., and A. H. Fitter. 1999. "The Magnitude and Control of Carbon Transfer between Plants Linked by a Common Mycorrhizal Network." *Journal of Experimental Botany* 50:9–13.

Rodriguez-Romero, Julio, Maren Hedtke, Christian Kastner, Sylvia Müller, and Reinhard Fischer. 2010. "Fungi, Hidden in Soil or Up in the Air: Light Makes a Difference." *Annual Review of Microbiology* 64:585–610.

Rolfe, R. T. and F. W. Rolfe. (1925) 2014. *The Romance of the Fungus World*. North Chelmsford, MA: Courier.

Roosth, Sophia. 2018. "Nineteen Hertz and Below: An Infrasonic History of the Twentieth Century." *Resilience: A Journal of the Environmental Humanities* 5, no. 3:109–24.

Rose, Deborah Bird. 2011. *Wild Dog Dreaming: Love and Extinction*. Charlottesville: University of Virginia Press.

Rosling, A., R. Landeweert, B. D. Lindahl, K. H. Larsson, T. W. Kuyper, A.F.S. Taylor, and R. D. Finlay. 2003. "Vertical Distribution of Ectomycorrhizal Fungal Taxa in a Podzol Soil Profile." *New Phytologist* 159, no. 3:775–83.

Roughgarden, Joan. 2013. *Evolution's Rainbow: Diversity, Gender, and Sexuality in Nature and People*. Berkeley: University of California Press.

Rowe, R. F. 1997. "The Commercial Harvesting of Wild Edible Mushrooms in the Pacific Northwest Region of the United States." *Mycologist* 11, no. 1:10–15.

Ruisi, Serena, Donatella Barreca, Laura Selbmann, Laura Zucconi, and Silvano Onofri. 2007. "Fungi in Antarctica." *Reviews in Environmental Science and Bio/Technology* 6, nos. 1–3:127–41.

Ruru, Jacinta. 2018. "Listening to Papatūānuku: A Call to Reform Water Law." *Journal of the Royal Society of New Zealand* 48, nos. 2–3:215–24.

Ryan, Frank. 2002. *Darwin's Blind Spot: Evolution beyond Natural Selection*. Boston: Houghton Mifflin Harcourt.

Saito, Haruo, and Gaku Mitsumata. 2008. "Bidding Customs and Habitat Improvement for Matsutake (*Tricholoma matsutake*) in Japan." *Economic Botany* 62, no. 3:257–68.

Sapp, Jan. 2004. "The Dynamics of Symbiosis: An Historical Overview." *Canadian Journal of Botany* 82, no. 8:1046–56.

Satsuka, Shiho. 2011. "Eating Others Well / Eating Well with Others." In "Poaching at the Multispecies Salon," special 100th issue, *Kroeber Anthropological Society Journal* 99/100:134–38.

———. 2013. "The Charisma of the Wild Mushroom." *RCC Perspectives*, no. 5:49–54.

———. 2018. "Sensing Multispecies Entanglements: Koto as an 'Ontology' of Living." *Social Analysis* 62, no. 4:78–101.

Sauer, Jonathan D. (1988) 1991. *Plant Migration: The Dynamics of Geographic Patterning in Seed Plant Species*. Berkeley: University of California Press.

Schilder, Annemiek. 2020. "'Tis the Time for Mushrooms." University of California Extension Service. https://ucanr.edu/blogs/blogcore/postdetail.cfm?postnum=41266.

Schwartz, Judith D. 2013. *Cows Save the Planet, and Other Improbable Ways of Restoring Soil to Heal the Earth*. White River Junction, VT: Chelsea Green.

Sebok, Thomas. 2009. "Biosemiotics: Its Roots, Proliferation and Prospects (2001)." In *Essential Readings in Biosemiotics*. Edited by Donald Favareau, 217–36. New York: Springer.

Sha, Tao, Han-Bo Zhang, Hua-Sun Ding, Zong-Ju Li, Li-Zhong Cheng, Zhi-Wei Zhao, and Ya-Ping Zhang. 2007. "Genetic Diversity of *Tricholoma matsutake* in Yunnan Province." *Chinese Science Bulletin* 52, no. 9:1212–16.

Shae, Shannon. 2018. "Behind the Scenes: How Fungi Make Nutrients Available to the World." *Office of Science, Department of Energy*, January 31. https://www.energy.gov/science/articles/behind-scenes-how-fungi-make-nutrients-available-world.

Shapiro, Kenneth. 2008. *Human-Animal Studies: Growing the Field, Applying the Field*. Ann Arbor, MI: Animals and Society Institute. https://www.animalsandsociety.org/public-policy/policy-papers/growing-the-field-applying-the-field/.

Sheldrake, Merlin. 2020. *Entangled Life: How Fungi Make Our Worlds, Change Our Minds and Shape Our Futures*. New York: Random House.

Shigo, Alex L. 1979. *Tree Decay: An Expanded Concept*. Washington, DC: Department of Agriculture, Forest Service.

Simard, Suzanne W., David A. Perry, Melanie D. Jones, David D. Myrold, Daniel M. Durall, and Randy Molina. 1997. "Net Transfer of Carbon between Ectomycorrhizal Tree Species in the Field." *Nature* 388, no. 6642:579–82.

Singer, Rolf. 1949. "The Agaricales in Modern Taxonomy." *Lilloa* 22:1–832.

Sivinski, John. 1981. "The Nature and Possible Functions of Luminescence in Coleoptera Larvae." *Coleopterists' Bulletin* 35, no. 2:167–79.

Skerratt, Lee Francis, Lee Berger, Richard Speare, Scott Cashins, Keith Raymond McDonald, Andrea Dawn Phillott, Harry Bryan Hines, and Nicole Kenyon. 2007. "Spread of Chytridiomycosis Has Caused the Rapid Global Decline and Extinction of Frogs." *EcoHealth* 4, no. 2:125–34.

Smil, Vaclav. 1980. "Environmental Degradation in China." *Asian Survey* 20, no. 8:777–88.

Smith, Sally E., and F. Andrew Smith. 2011. "Roles of Arbuscular Mycorrhizas in Plant Nutrition and Growth: New Paradigms from Cellular to Ecosystem Scales." *Annual Review of Plant Biology* 62:227–50.

Smith, S. E., and D. Read. 2008. *Mycorrhizal Symbiosis* 3rd ed. New York: Academic.

Söderström, B. E., and D. J. Read. 1987. "Respiratory Activity of Intact and Excised Ectomycorrhizal Mycelial Systems Growing in Unsterilized Soil." *Soil Biology and Biochemistry* 11:231–37.

Spencer, Herbert. 1864. *Principles of Biology*. London: Williams and Norgate.

Stamets, Paul. 2005. *Mycelium Running: How Mushrooms Can Help Save the World*. New York: Random House Digital.

Steinhardt, Joanna. 2018. "Psychedelic Naturalism and Interspecies Alliance: Views from the Emerging Do-It-Yourself Mycology Movement." In *Plant Medicines, Healing and Psychedelic Science*, edited by B. C. Labate and C. Cavnar, 167–84. New York: Springer.

Sterelny, Kim. 2005. "Made by Each Other: Organisms and Their Environment." *Biology and Philosophy* 20, no. 1:21–36.

Sterne, Jonathan. 2003. *The Audible Past*. Durham, NC: Duke University Press.

Stone, Jeffrey K., Charles W. Bacon, and James F. White Jr. 2000. "An Overview of Endophytic Microbes: Endophytism Defined." In *Microbial Endophytes*, edited by C. W. Bacon and J. F. White Jr., 17–44. Boca Raton, FL: CRC.

Stone, Jeffrey K., Jon D. Polishook, and James F. White. 2004. "Endophytic Fungi." In *Biodiversity of Fungi*, edited by Greg Mueller, Mercedes Foster and Gerald Bills, 241–70. Burlington, ON: Elsevier Academic.

Strobel, Gary, and Bryn Daisy. 2003. "Bioprospecting for Microbial Endophytes and Their Natural Products." *Microbiology and Molecular Biology Review* 67, no. 4:491–502. https://doi.org/10.1128/MMBR.67.4.491-502.2003.

Subramaniam, Banu. 2001. "The Aliens Have Landed! Reflections on the Rhetoric of Biological Invasions." *Meridians: Feminism, Race, Transnationalism* 2, no. 1:26–40.

Summerbell, Richard C. 2005. "From Lamarckian Fertilizers to Fungal Castles: Recapturing the Pre-1985 Literature on Endophytic and Saprotrophic Fungi Associated with Ectomycorrhizal Root Systems." *Studies in Mycology* 53:191–256.

Sutton, Brian. 1996. *A Century of Mycology*. Cambridge: Cambridge University Press.

Talbot, J. M., S. D. Allison, and K. K. Treseder. 2008. "Decomposers in Disguise: Mycorrhizal Fungi as Regulators of Soil C Dynamics in Ecosystems under Global Change." *Functional Ecology* 22, no. 6:955–63. https://www.bu.edu/articles/2016/soil-fungi/.

TallBear, Kim. 2011. "Why Interspecies Thinking Needs Indigenous Standpoints." *Fieldsights*, November 18. Society for Cultural Anthropology. https://culanth.org/fieldsights/why-interspecies-thinking-needs-indigenous-standpoints.

Tashi, Nyima, Tang Yawei, and Zeng Xingquan. 2013. "Food Preparation from Hulless Barley in Tibet." In *Advance in Barley Sciences: Proceedings of 11th International Barley Genetics Symposium*, edited by G. Zhang, C. Li, and X. Liu, 151–58. New York: Springer.

Thompkins, Peter, and Christopher Bird. 1973. *The Secret Life of Plants*. New York: Harper and Row.

Todd, Zoe. 2014. "Fish Pluralities: Human-Animal Relations and Sites of Engagement in Paulatuuq, Arctic Canada." *Études/Inuit/Studies* 38, nos. 1–2:217–38.

———. 2016. "An Indigenous Feminist's Take on the Ontological Turn: 'Ontology' Is Just Another Word for Colonialism." *Journal of Historical Sociology* 29, no. 1:4–22.

Tominaga, Yasuto, Ryoko Arai, and Toshio Ito. 1981. "Matsutake of the People's Republic of China, 1: Matsutake of Yunnan Province." *Hiroshima Agricultural College Bulletin* 6, no. 4:449–58.

Tønnessen, Morten, Timo Maran, and Alexei Sharov. 2018. "Phenomenology and Biosemiotics." *Biosemiotics* 11, no. 3:323–30.

Trappe, James M. 1977. "Selection of Fungi for Ectomycorrhizal Inoculation in Nurseries." *Annual Review of Phytopathology* 15, no. 1:203–22.

———. 2005. "A. B. Frank and Mycorrhizae: The Challenge to Evolutionary and Ecologic Theory." *Mycorrhiza* 15, no. 4:277–81.

Trudell, S. A., J. Xu, I. Justo, A. Saar, and J. Cifuentes. 2017. "North American Matsutake: Names Clarified and a New Species Described." *Mycologia* 109:379–90.

Tsing, Anna Lowenhaupt. 2012. "Unruly Edges: Mushrooms as Companion Species." *Environmental Humanities* 1, no. 1:141–54.

———. 2015. *The Mushroom at the End of the World: On the Possibility of Life in Capitalist Ruins.* Princeton, NJ: Princeton University Press.

———. 2018. "A Multispecies Ontological Turn?" In *The World Multiple: The Quotidian Politics of Knowing and Generating Entangled Worlds,* edited by K. Omura, G. J. Ostuki, S. Satsuka, and A. Morita, 233–47. New York: Routledge.

Tsing, Anna Lowenhaupt, Nils Bubandt, Elaine Gan, and Heather Anne Swanson, eds. 2017. *Arts of Living on a Damaged Planet: Ghosts and Monsters of the Anthropocene.* Minneapolis: University of Minnesota Press.

Tsing, Anna, and Shiho Satsuka. 2008. "Diverging Understandings of Forest Management in Matsutake Science." *Economic Botany* 62, no. 3:244–53.

Turner, J. Scott. 2009. *The Extended Organism: The Physiology of Animal-Built Structures.* Cambridge, MA: Harvard University Press.

Urry, Lisa, Michael Cain, Steven Wasserman, Peter Minorsky, and Jane Reece. 2016. *Campbell Biology.* New York: Pearson.

Uexküll, Jakob von. 1909. *Umwelt und Innenwelt der Tiere.* Berlin, Germany: J. Springer.

———. 2010. *A Foray into the Worlds of Animals and Humans: With a Theory of Meaning.* Translated by Joseph D. O'Neil. Minneapolis: University of Minnesota Press.

Vajda, Vivi, and Stephen McLoughlin. 2004. "Fungal Proliferation at the Cretaceous-Tertiary Boundary." *Science* 303, no. 5663:1489.

Van der Heijden, M., F. M. Martin, M. Selosse, and I. R. Sanders. 2015. "Mycorrhizal Ecology and Evolution: The Past, the Present, and the Future." *New Phytologist* 205, no. 4:1406–23.

van Dooren, Thom. 2019. *The Wake of Crows: Living and Dying in Shared Worlds.* New York: Columbia University Press.

van Dooren, Thom, Eben Kirksey, and Ursula Münster. 2016. "Multispecies Studies Cultivating Arts of Attentiveness." *Environmental Humanities* 8, no. 1:1–23.

van Dooren, Thom, and Deborah Bird Rose. 2016. "Lively Ethnography: Storying Animist Worlds." *Environmental Humanities* 8, no. 1:77–94.

Van Doremalen, Neeltje, Trenton Bushmaker, Dylan H. Morris, Myndi G. Holbrook, Amandine Gamble, Brandi N. Williamson, Azaibi Tamin, Jennifer L. Harcourt, Natalie J. Thornburg, and Susan I. Gerber. 2020. "Aerosol and Surface Stability of SARS-CoV-2 as Compared with SARS-CoV-1." *New England Journal of Medicine* 382, no. 16:1–4.

Varma, Ajit, Ram Prasad, and Narendra Tuteja, eds. 2017. *Mycorrhiza-Nutrient Uptake, Biocontrol, Ecorestoration.* New York: Springer.

Veits, Marine, Itzhak Khait, Uri Obolski, Eyal Zinger, Arjan Boonman, Aya Goldshtein, Kfir Saban, Rya Seltzer, Udi Ben-Dor, and Paz Estlein. 2019. "Flowers Respond to Pollinator Sound within Minutes by Increasing Nectar Sugar Concentration." *Ecology Letters* 22, no. 9:1483–92.

Voitk, Aare, John Maunder, and Andrus Votik. 2012. "Slugs and Mushrooms." *Omphalina: Newsletter of Foray Newfoundland and Labrador* 3, no. 6:4–7.

Volk, Tom. 2000. "*Schizophyllum commune*, the Split Gill Fungus, Perhaps the World's Most Widespread Fungus—and Possessor of Over 28,000 Different Sexes." *Tom Volk's Fungus of the Month for February.* https://botit.botany.wisc.edu/toms_fungi/feb2000.html.

Volk, Tyler. 2007. "The Properties of Organisms Are Not Tunable Parameters Selected Because They Create Maximum Entropy Production on the Biosphere Scale: A By-Product Framework in Response to Kleidon." *Climatic Change* 85, nos. 3–4:251–58.

Volkov, Alexander G., and Yuri B. Shtessel. 2020. "Underground Electrotonic Signal Transmission between Plants." *Communicative and Integrative Biology* 13, no. 1:54–58.

von Frisch, Karl. (1927) 1953. *The Dancing Bees*. London: Methuen.

Walsh, Patrick T., Mike Hansell, Wendy D. Borello, and Susan D. Healy. 2011. "Individuality in Nest Building: Do Southern Masked Weaver (*Ploceus velatus*) Males Vary in Their Nest-Building Behaviour?" *Behavioural Processes* 88, no. 1:1–6.

Wang, Chenglei. 2014. "Do Not Ignore the Role of Fungi in Biodiversity Conservation." *Biodiversity Conservation*, March 23. https://biodiversityconservationblog.wordpress.com/2014/05/23/do-not-ignore-the-role-of-fungi-in-biodiversity-conservation/.

Wang, Yun, Chunxiang Zhang, and Shuhong Li. 2017. "*Tricholoma matsutake*: An Edible Mycorrhizal Mushroom of High Socioeconomic Relevance in China." *Revista Mexicana de Micología* 46:55–61.

Ward, Peter, Conrad Labandeira, Michel Laurin, and Robert A. Berner. 2006. "Confirmation of Romer's Gap as a Low Oxygen Interval Constraining the Timing of Initial Arthropod and Vertebrate Terrestrialization." *Proceedings of the National Academy of Sciences* 103, no. 45:16818–22.

Watling, Roy. 1980. *How to Identify Mushrooms to Genus V: Cultural and Developmental Features*. Eureka, CA: Mad River.

Weisman, Alan. 2007. *The World without Us*. New York: Picador.

Wilkening, Kenneth E. 2004. *Acid Rain Science and Politics in Japan: A History of Knowledge and Action toward Sustainability*. Cambridge, MA: MIT Press.

Wilkinson, David M. 2006. *Fundamental Processes in Ecology: An Earth Systems Approach*. Oxford: Oxford University Press.

Winkler, Daniel. 1998. "Deforestation in Eastern Tibet: Human Impact—Past and Present; Development, Society and Environment in Tibet." *Proceedings of 7th Seminar of the International Association of Tibetan Studies (IATS) 1995, Austria, Vienna*, 79–96.

———. 2008. "The Mushrooming Fungi Market in Tibet Exemplified by *Cordyceps sinensis* and *Tricholoma matsutake*." *Journal of the International Association of Tibetan Studies* 4:1–47.

———. 2008. "Yartsa Gunbu (*Cordyceps sinensis*) and the Fungal Commodification of Tibet's Rural Economy." *Economic Botany* 62, no. 3:291–305.

Witzany, Günther. 2006. "Plant Communication from Biosemiotic Perspective: Differences in Abiotic and Biotic Signal Perception Determine Content Arrangement of Response Behavior. Context Determines Meaning of Meta-, Inter- and Intraorganismic Plant Signaling." *Plant Signaling and Behavior* 1, no. 4:169–78.

———, ed. 2012. *Biocommunication of Fungi*. New York: Springer Science and Business Media.

Witzany, Günther, and František Baluška, eds. 2012. *Biocommunication of Plants*. Vol. 14. New York: Springer Science and Business Media.

Wohlleben, Peter. 2015. *The Hidden Life of Trees: What They Feel, How They Communicate—Discoveries from a Secret World*. Vancouver, BC: Greystone Books.

———. n.d. "The Hidden Life of Trees." *Canadian Science Publishing*. Accessed April 30, 2020. http://blog.cdnsciencepub.com/the-hidden-life-of-trees-by-peter-wohlleben/.

Wolf, Eric R. (1982) 2010. *Europe and the People without History.* Berkeley: University of California Press.

Wright, Justin P., and Clive G. Jones. 2006. "The Concept of Organisms as Ecosystem Engineers Ten Years On: Progress, Limitations, and Challenges." *BioScience* 56, no. 3:203–9.

Wu, Keliang, and Wu Changxin. 2004. "Documentation and Mining of Yak Culture to Promote a Sustainable Yak Husbandry." International Congress on Yak, Chengdu, Sichuan, PR China, 2004. Proceedings, ed. J. Zhong, X. Zi, J. Han, Z. Chen, International Veterinary Information Service, Ithaca, NY. www.ivis.org.

Wynne, Clive D. 2007. "What Are Animals? Why Anthropomorphism Is Still Not a Scientific Approach to Behavior." *Comparative Cognition and Behavior Reviews* 2:125–35.

Xu, Jianchu, and Andreas Wilkes. 2004. "Biodiversity Impact Analysis in Northwest Yunnan, Southwest China." *Biodiversity and Conservation* 13, no. 5:959–83.

———. 2004. "People and Ecosystems in Mountain Landscape of Northwest Yunnan, Southwest China: Causes of Biodiversity Loss and Ecosystem Degradation." *Global Environmental Research—English Edition* 6, no. 1:103–10.

Xu, Junqian. 2017. "The Mushrooming Demand for Mushrooms." *China Daily,* August 26. http://www.chinadaily.com.cn/a/201708/26/WS59bb7040a310d4d9ab7e7bff_3.html.

Xue, Hui. 2006. "Assessing the Role of Risk in the Agro-pastoral Systems of Northwest Yunnan Province, China." Master's thesis, School of Resource and Environmental Management, Simon Fraser University, Burnaby, BC.

Yamanaka, Takashi, Akiyoshi Yamada, and Hitoshi Furukawa. 2020. "Advances in the Cultivation of the Highly-Prized Ectomycorrhizal Mushroom *Tricholoma matsutake.*" *Mycoscience* 61, no. 2:49–57.

Yang, Bin. 2004. "Horses, Silver, and Cowries: Yunnan in Global Perspective." *Journal of World History* 15, no. 3:281–322.

———. 2008. *Between Winds and Clouds: The Making of Yunnan.* New York: Columbia University Press.

Yang, Xuefei, Jun He, Chun Li, Jianzhong Ma, Yongping Yang, and Jianchu Xu. 2006. "Management of Matsutake in NW-Yunnan and Key Issues for Its Sustainable Utilization." In *Sino-German Symposium on the Sustainable Harvest of Non-timber Forest Products in China,* edited by C. Kleinn, Y. Yang, H. Weyerhaeuser, and M. Stark, 48–57. Göttingen, Germany: World Agroforestry Centre.

Yang, Yu-Hua, Ting-You Shi, Yong-Shun Bai, Kai-Mei Su, Hong-Fen Bai, Li-Qiong Mu, Yan Yu, Xing-Zhou Duan, Zheng-Jun Liu, and Chun-De Zhang. 2007. "Discussion on Management Model of Contracting Mountain and Forest about Bio-resource Utilization under Natural Forest in Chuxiong Prefecture." *Forest Inventory and Planning* 3:87–89.

Yeh, Emily T. 2000. "Forest Claims, Conflicts and Commodification: The Political Ecology of Tibetan Mushroom-Harvesting Villages in Yunnan Province, China." *China Quarterly,* no. 161:264–78.

———. 2013. "Blazing Pelts and Burning Passions: Nationalism, Cultural Politics, and Spectacular Decommodification in Tibet." *Journal of Asian Studies* 72, no. 2:319–44.

Yeh, Emily, and Chris Coggins, eds. 2014. *Mapping Shangrila: Contested Landscapes of the Sino-Tibetan Borderlands.* Seattle: University of Washington Press.

Yin, Xiulian, Qinghong You, and Zhonghai Jiang. 2012. "Immunomodulatory Activities of Different Solvent Extracts from *Tricholoma matsutake* (S. Ito et S. Imai) Singer (Higher Basidiomycetes) on Normal Mice." *International Journal of Medicinal Mushrooms* 14, no. 6:549–56.

Yong, Ed. 2015. "Why Do Most Languages Have So Few Words for Smells?" *Atlantic*, November 6. https://www.theatlantic.com/science/archive/2015/11/the-vocabulary-of-smell/414618/.

———. 2019. "The Overlooked Organisms That Keep Challenging Our Assumptions about Life." *Atlantic*, January 17. https://www.theatlantic.com/science/archive/2019/01/how-lichens-explain-and-re-explain-world/580681/.

Yu, Dan Smyer. 2013. *The Spread of Tibetan Buddhism in China: Charisma, Money, Enlightenment.* New York: Routledge.

Yu, Zhenzhong, and Reinhard Fischer. 2019. "Light Sensing and Responses in Fungi." *Nature Reviews Microbiology* 17, no. 1:25–36.

Zeder, Melinda A. 2015. "Core Questions in Domestication Research." *Proceedings of the National Academy of Sciences* 112, no. 11:3191–98.

Zeller, Sanford Myron, and K. Togashi. 1934. "The American and Japanese Matsu-takes." *Mycologia* 26, no. 6:544–58.

Zhong, Qiu, and Guoqing Shi. 2020. "Environmental Issues to Protection: The United States versus China." In *Environmental Consciousness in China*, 11–31. Cambridge, MA: Chandos.

Zimmerman, Erin. 2017. "*The Hidden Life of Trees*, by Peter Wohlleben." *Canadian Science Publishing*, November 13. http://blog.cdnsciencepub.com/the-hidden-life-of-trees-by-peter-wohlleben/.

Zhou, Yongming. 1999. *Anti-drug Crusades in Twentieth-Century China: Nationalism, History, and State Building.* Lanham, MD: Rowman and Littlefield.

Zhu, Yanping, Jieru Pan, Junzhi Qiu, and Xiong Guan. 2008. "Isolation and Characterization of a Chitinase Gene from Entomopathogenic Fungus Verticillium Lecanii." *Brazilian Journal of Microbiology* 39, no. 2:314–20.

ACKNOWLEDGMENTS

First and foremost, I would like to acknowledge my great debt and gratitude to the other members of the Matsutake Worlds Research Group (Tim Choy, Lieba Faier, Elaine Gan, Miyako Inoue, Shiho Satsuka, and Anna Tsing), for creating such an inspiring and supportive group of fellow scholars and adventurers. I owe a special thanks to Shiho, for her many insights and questions that open the mind. In my intellectual life, Anna has been my biggest inspiration and best interlocutor, shaping my work since we met when I was an undergraduate. We were all brought together through Anna's networks, and I will always be grateful for the connections we made to such world-changing thinkers such as Donna Haraway and Margaret McFall-Ngai.

At Michigan, a number of people inspired me, including Gillian Feely-Harnik and her work on beavers, Tom Trautmann and his work on the history of elephant-human relations, and Erik Mueggler and his incredible scholarship on Yi history and worlds. Arun Agrawal remains a guiding light of mentorship and scholarship that crosses disciplinary divides. In California, my good friends Ellen Baker and Freddy Menge helped bring me into these mushroom worlds, always cooking up great food and wonderful conversation. My cousins Gordy Slack and Adrianna Taranta have nurtured me through the years and shared many excitements.

At Simon Fraser University, where I had the good fortune and privilege to teach for the past thirteen years, I would like to thank the members of my reading group, Travers, Lindsey Freeman, Jessi Lee Jackson, Amanda Watson, Nick Scott, Stacy Pigg, Coleman Nye, and especially Kathleen Millar. My department colleague Sonja Luehrmann, who left this world much too soon, provided her insights into this project. Susan

Erikson and Nicole Berry have always been there for encouragement and excellent suggestions. I would also like to offer a special thanks to some of my remarkable students: Morgaine Lee, Cheyanne Connell, Kathleen Inglis, and Jelena Golubović, whose reflections and assistance were of great importance.

On the mycology front, I am grateful to Willoughby Arevalo, Shannon Berch, and Zaac Chaves, each of whom read the manuscript, saving me from a number of mistakes and offering many helpful suggestions. Mendel Skulski's deep curiosity into fungal capacities proved inspiring, and Oron Frenkel's thinking on the limits of the human has pushed my own. I am indebted to Rob Dunn, a scientific renaissance man who is also a fantastic popular writer, for his astute guidance in helping me improve the book as a reader for Princeton.

Thanks to James C. Scott and Kalyanakrishnan (Shivi) Sivaramakrishnan for inviting me to speak at the Agrarian Studies Seminar in the spring of 2014 and to Michael Dove for inviting me back to Yale that fall. I would like to thank Karen Hébert for hosting me there and being a wonderful friend over the years. She joins other graduate school friends who nourished me with their love and great intelligence, including Edward Murphy, Genese Sodikoff, and Mel Johnson, who read many drafts and provided sage advice.

At the University of Washington, I greatly appreciated the conversations with Celia Lowe, Gaymon L. Bennett, Luke Bergman, and Stevan Harrell. At the University of British Columbia, I relished talks with Julie Cruickshank, Tim Cheek, Mark Werner, and Jack Patrick Hayes. At Whittier College, I was hosted by Robert Marks, whose work has enriched me for decades.

In Australia, where I initially traveled to talk about wild elephants in China, conversations eventually expanded into mushrooms, and I made new lifelong friends including Matt Chrulew, Thom van Dooren, and Fiona Probyn-Rapsey

A seminar at the Rachel Carson Center in Munich, Germany, proved to be especially important in my thinking, particularly through meeting Ursula Munster, Katherine Gibson, Eben Kirksey, and Myra Hird. At Oxford University, Anna Lora-Wainwright and Loretta Lou raised a number of fascinating questions and points. At Nipissing University, I benefited by thinking aloud with Dean Bavington and Jennifer Lee Johnson. Many thanks to Brett Buchanan for inviting me to his field philosophy workshop in Paris, where I met Vinciane Despret, whose work has deeply influenced my own. A semester at Hawai'i, where I was lucky enough to have Cathy Clayton, Anthony Amend, and Jonathan Padwe around for many discussions, helped expand the book into new directions. Berkeley maintained its reputation as a place of inquisitive intellectualism, as exemplified by my interactions with You-tien Hsing, Jerry Zee, and Pheng Cheah.

In China, many friends and colleagues assisted me: Xu Jianchu, Yu Fuqiang, Su Kaimei, Yang Xueqing, Luo Wenhong, Zhai Wen, Yang Haiyu, Pai Yi, Zhang Hai, Cui Yi, and Wang Yun. In Japan, I offer a hearty thanks to Dr. Yoshimura. The project was enriched by my presentations at Kyoto University, with a special thanks to Noboru Ishikawa. I have now enjoyed many discussions with Jun Akamine, from Hitotsubashi University, who engaged our matsutake project with much gusto and careful thinking.

Generous funding was provided by the Social Sciences and Humanities Research Council, the Social Science Research Council's Transregional Research Fellowship, the American Council of Learned Societies and Luce Foundation Fellowship, Simon Fraser University's Publication Fund and the Toyota Foundation. Some of this material was previously published in my article in *Social Analysis*.

Over the years, Kathy White has provided wonderfully astute editorial guidance, often with great wit and insight. She was essential in my first book and remains so for my second.

I am grateful to my parents, Walton and Peggy, for all that they have done to raise me with such love and support. Their deep love for the natural world—paddling out to see the moose, learning bird songs, pointing out the flowers—has inspired my own way of being.

Of all these people, my main source of daily inspiration and liveliness is my partner, Leslie Walker Williams. She has talked more about this amazing journey with me and read more drafts than I dare remember or repeat. Even when I first met Leslie more than thirty years ago, she surprised me with her intense suspicion of anthropocentrism, matched by her strong interest in the liveliness of all beings. These sensibilities might have guided her renowned morel mushroom hunting capacities and her great abilities as a beekeeper. Leslie, together with my children, Walker and Logan, bring me joy and keep me from being pulled completely into fungal networks, keeping my head above the ground.

INDEX

Page numbers in italics refer to illustrations

actants (Actor Network Theory), 19, 189–90, 216n38

active niche theory, 62–63, 72–73, 220n25, 221nn28–29

Agamben, Giorgio, 86–87

Agaricus mushroom, 10, 124; *Agaricus ponderosus/magnivelaris*, 215n28

agency, 18–24, 61–62, 71–73, 201–2; cumulative agency, 50–51. *See also* language and terminology; world-making, concept of

Alice in Wonderland mushroom, 10

Allotropa (candy cane plant), 30, 39, 109, 111–12, 148

Amanita, 124, 205; *Amanita citrina* and *Amanita muscaria*, 10; *Amanita phalloides* (death cap), 227n9

animals and mammals: anthropomorphism and, 226n79; biology's animal-centered view, 31, 71–72, 78, 217n3; fungal communication with, 95–96, 227n10; in history of life, 43; language and animacy of animals, 52–55; non-human characters in literature, 213–14n3; plant communications and, 89; relationships among organisms, 157–58, 233n20; spore dispersal role, 44–45, 165–66; in taxonomy of who eats whom, 64–66. *See also* biology; human exceptionalism and anthropomorphism; perceptions, animal-centric view of

anthropological research: attending to human difference, 83–84; author's

interest and approach, xiii, xvi–xvii, 15–18, 51–52, 193

anthropomorphism. *See* human exceptionalism and anthropomorphism

antibiotics, 44, 219n38

anymal, 4–5

Approaches to Plant Evolutionary Ecology (Cheplick), 64

Arai, Ryoko, 117–18

arbuscular mycorrhizae formations (AM), 38. *See also* mycorrhizae and mycorrhizal relationships

Arevalo, Willoughby, 197, 215n29

Armillaria fungus, 105, 205, 215n28; largest living organism, 8

Arora, David, 109, 173–74

artist's conk fungus (*Ganoderma applanatum*), 188

Attenborough, David, 31, 217n5

Bai people in Yunnan, 175–76. *See also* Yunnan Province

Barad, Karen, xi

barley fungus: *Claviceps* and *Fusarium* sp., 160

barley world-making, 158–61, *161*, 169–70, 184; impact of matsutake on, 164–65

bats and fungal white-nose syndrome, 43

Beauveria bassiana, 96–97, 225n70

beetles, 46; mountain pine beetle (*Dendroctonus ponderosae*), 188; truffle beetle (*Leoides*), 95–96

Japanese red pine (*Pinus densiflora*), 229n33;
matsutake partner, 107–8
Johnson, Melissa: *Becoming Creole*, 158
Jones, Clive G., 65
Jongmans, Antoine G., 35–36

Kant, Immanuel, 84
Kimmerer, Robin Wall, 52–54, 188
kingdom of fungi. *See* fungi, kingdom of
Kish, Daniel, 85
Kohn, Eduard, 216n45
Kunming Institute of Botany (KIB), 115

laccaris bicolor, 32
Lactarius hibbardae, 10
Landecker, Hannah, 219n38
language and terminology: "agency," use of,
18, 50–51, 61–62, 71–73, 201–2; animacy of
animals, plants, and fungi in, 52–55; biology's mechanistic approach, 3–7, 14–15,
17, 220n23; "breathing," use of, 27, 216n53;
colonialism and biology, 215n29; "ecosystem
engineering," use of, 233n21; "fruiting"
terminology, 166; "growing," use of, 64,
221n32; human exceptionalism, 4–5, 216n53;
"infection," use of, 58–59; "learning"
and "decide," use of, 223n14, 226n80; in
microbiology, 34, 58–59; "moment,"
definition and human umwelt, 82–83;
"ontology," use of, 20; passive portrayals,
35–36, 51–52, 54, 61–62, 65; plant-based
terms for fungi, 226n75; senses shaping,
84, 224n36; shaping what we know, 55,
56–61; Yi language in matsutake economy,
136–38, 141–42. *See also* biosemiotics;
world-making, concept of
Lao Wu (Old Wu, matsutake hunter), 146–47
laoban (bosses in Yunnan), xvi, 130–38,
180–82, 231n7. *See also* Yunnan Province
Latour, Bruno, 19, 189–90, 216n38, 216n39
Leary, Timothy, 219n5
Lévi-Strauss, Claude, 198
Lewes, George Herbert, 226n79
Lhasa, 178, 179

Li, Bo, *129*, 129–30
lichens: cushioning matsutake, 2; fungi's
variety and, *xiv*, 186–87; growing on tombstones, 9, *9*; story of a lichen, 202–3; symbiotic relations, 192; in urban settings, 193–95
"life partners": "critical life partners," 200.
See also mycorrhizae and mycorrhizal
relationships; trees and forests; world-
making, concept of
lignin, 39–41. *See also* trees and forests
Li Laoban (Boss Li, matsutake dealer),
130–36, 137–38; fenced-in matsutake
patch, 130–31, 140
Li, Mingwo (matsutake hunter), 145, 147
limestone, 218n26
Linné, Carol von, 215n28
literature and literary traditions: non-human
characters in, 213–14n3
lively approach: world-making concept as,
190. *See also* world-making, concept of
Lorimer, Jamie, 83
Luo, Wenhong, 127

Mackinnon, Andy, 93
magic mushrooms: origin of term, 219n5
Majid, Asifa, 84
mammals. *See* animals and mammals
Mandarin, 136–37, 141
Mao Zedong era: Yi economy during, 128,
133–34, 139–40; Yunnan Tibetan economy
during, 156. *See also* Yunnan Province
maple trees, 194
matsutake economy, 103–4; compared to
other mushrooms, 132–33, *133*, 171–72;
cultivation and, 166, 201–2; demand for
freshness, 120–21, 145, 180; ethnography of
Yi's, 125–30; gender roles in, 176–77; Japan
and, xv–xvi, 116–17, 172, *173*, 183, 229n36,
230n37; laws effecting, 122, 151, 184, 230n46;
matsutake money and yak money, 15,
176–77, 185; mushroom gold, 118, 173, 183;
non-human actors in, 122–23; pickled
(brined) matsutake, 119; power of matsutake
dealers, 134–36, *135*; prices, 118–20, 183–84;